PASCAL for Electronics and Communications

Richard Meadows

BSc, MSc, PhD, FIEE, CEng, FIElecIE, MInstP, ARCS
Head of Electronic and Communications Engineering
The Polytechnic of North London

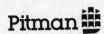

PITMAN PUBLISHING
128 Long Acre, London WC2E 9AN

A Division of Longman Group UK Limited

© R Meadows 1985

First published in Great Britain 1985
Reprinted 1987, 1988, 1989

ISBN 0 273 02155 9

Printed and bound in Singapore

Contents

Preface

This book has been written to act as a practical guide to learning how to write programs in *Standard* PASCAL. No prior knowledge of computing or the PASCAL language is assumed, nor is it needed to follow the text.

The aim of the book is to provide students and engineers new to PASCAL with the knowledge and understanding of this powerful language so that they can construct their own programs to solve a wide variety of problems.

The text follows a logical sequence and at all stages the concepts being introduced are backed by practical program examples drawn principally from basic electronics, circuit theory and communications. Exercise problems are also given at the end of each Chapter (other than the first introductory one) and typical program solutions for these are given in the answer section at the end of the book.

Finally, I should like to express my thanks to my colleagues Mike Brinson and Larry Parsons for their many helpful discussions and comments and as always my Editor, John Cushion, for his continued support.

Richard Meadows
April 1985

Dedication
To my wife Lynn and students at PNL and City College.

1 Introduction

1.1 Introduction: computer hardware and software

Before starting PASCAL, this chapter briefly reviews some of the important terms associated with computers and programming and at the same time gives a simple insight into how a digital computer operates to "translate" and "run" programs, whether they are written in PASCAL or some other computer language.

HARDWARE: the physical components of a computer system

Hardware is used to describe any component or unit used in the construction of a digital computer system. Input devices, microprocessor chips, memory chips, I/O chips, auxiliary stores, output devices are all examples of hardware.

Fig. 1.1 shows a block diagram of the basic units—the hardware—associated with a typical computer system.

The function of these basic units is as follows:

Input devices Input devices are required to input data, instructions, programs, etc. to the central processing unit of the computer. They enable the user to communicate with computers. Examples of input devices include:

Keyboard devices (similar to electric typewriters).

Transducers plus analog-to-digital (A-to-D) converters to convert the transducer signals, e.g. electrical signals from temperature, light, pressure sensors, to the digital binary-coded signals that the computer can "understand".

Tape recorders or cassette players containing programs, data, etc. in digital form stored on magnetic tape.

Disc drive units—more readily accessible from the computer's point of view and therefore much speedier than cassettes—where programs, etc. are stored on magnetic discs.

Output devices Output devices are required to translate the computer output, which is in binary form, to human-recognisable form or a form suitable for controlling a machine, etc.. Examples of output devices include:

Printers (to provide "hardcopy" of results).

Visual display units (normally abbreviated to VDU) which display results, graphics, games, etc. on a tv-type screen).

Lamps

Light-emitting diodes (LEDs).

Liquid-crystal displays (LCD).

Speech devices.

Digital-to-analog (D-to-A) converters for converting the computer output to an analog or continuous-type signal form, which, for example, may then be amplified and used to control a specific function in a process or machine.

Input/output unit (I/O unit)   The input/output (normally abbreviated to I/O) unit forms the interface between the computer and the input and output devices. The I/O unit feeds information in the correct binary-coded form and sequence into the central processor unit from input devices, and also outputs results from the computer to output devices.

The input and output lines may enter a common I/O unit which may be programmed to control both the input and output data flow to and from the central processor; or, alternatively, the I/O unit function may be split into two separate devices. The term *input port* is used to denote both the entry point and/or the device handling input data; the term *output port* to denote the output point/device for output data.

Peripheral devices   All devices, such as input and output devices, connected to the I/O ports are known as peripherals.

The central processing unit (CPU)   The actual computer, as distinct from its peripheral devices, is known as the central processing unit, the CPU. The CPU consists essentially of three basic units: a *memory* unit, an *arithmetic-logic* unit, and a *control* unit. The operation of these three sub-units making up the CPU are closely interrelated. In many microcomputers all three may be contained in as few as two or three or even a single integrated circuit chip. The functions of the three CPU units are as follows:

The store or memory unit   The store or memory holds the data and instructions of the program which have been fed into the CPU via the I/O unit. It also holds the permanently stored programs (*operating systems programs*) that permit the operation of the computer system immediately after switch-on.

The arithmetic and logic unit (ALU)   The ALU performs arithmetic operations ($+ - \times \div$) and logic decisions on the data fed to it from the memory unit, according to the instructions of the program.

The control unit (CU)   The control unit interprets and carries out the instructions of the program in the exact sequence—instruction by

Fig. 1.1 Hardware associated with a typical digital computer system

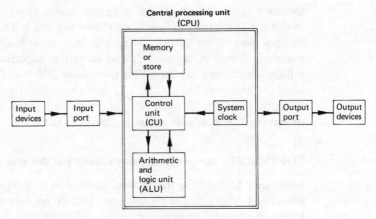

(a)   Block diagram of a fundamental digital computer system

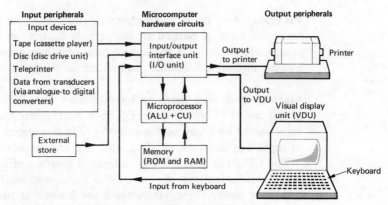

(b)   A block schematic showing units making up a microcomputer system plus some typical peripherals

instruction—as stated by the program held in the memory unit. The CU is the master unit controlling the processing of all instructions and the movement of data to and from the input/output devices to the memory unit and to and from the ALU.

The clock   A digital computer is the prime example of a digital-logic sequential system and as such requires a very accurate timing source to ensure perfect synchronism of all its step-by-step operations. The timing source controlling these operations in the CPU is known as the clock. The clock invariably consists of a crystal-controlled oscillator circuit which generates a continuous wavetrain of rectangular pulses of very stable frequency. These pulses are fed to all circuits in the computer system to ensure that they work in exact synchronism.

Auxiliary storage devices   In many applications the internal store or memory unit of the CPU will prove inadequate. There will not be enough memory space to store *all* program instructions, data, results, etc. To overcome this problem, auxiliary stores are used. These consist of magnetic devices where programs, data, etc. are stored on magnetic tapes and magnetic discs. A tape of 100 m length is capable of storing about one million characters, whilst a disc may store 200+ million. Information from auxiliary stores is fed via the I/O unit to the internal store when it is called upon by the CPU. It must, of course, be split into suitable blocks to avoid swamping the available memory space in the CPU.

SOFTWARE: the program instructions for the computer system

Software Software is the general term used to denote all forms of program associated with computing systems, just as hardware is used to denote all forms of physical component.

Here are some important software terms:

Program Any list of instructions or routines or actions set out in a logical order to solve a particular problem is known as a program. All digital computers work on a sequential basis. Each instruction is carried out one by one until all the instructions making up the program have been executed.

Machine code and machine instructions The CPU itself works with programs coded in binary form, i.e. each instruction and every piece of data is represented by a unique pattern of 1–0 binary digits. The binary codes used to represent instructions, data, etc., are known as machine codes.

The CPU can perform only a limited number of instructions, the number being determined in the case of a microcomputer by the actual microprocessor used. These instructions are known as machine instructions and the complete set available for a given microprocessor is known as its *instruction set*.

Object program A list of machine instructions set up in a logical order to solve a given problem is known as the object program.

Source programs Machine code programs (i.e. object programs) are very tedious to compose directly since they consist entirely of binary 1–0 patterns. Programming is normally done using one of the following two language forms:

a) High-level Languages
Such as PASCAL, BASIC, ALGOL, FORTRAN, etc.
These languages are very much easier for humans to understand and use. However, the computer cannot work directly with programs composed in a high-level language. Such programs must first be translated into machine code before the CPU can begin its work.

Programs written in a high-level language are known as **source programs**. A program which translates the source program into the machine code program (the object program) for execution by the computer is known as a **compiler**.

b) *Symbolic or Assembly Codes*

Symbolic or assembly code programming is essentially a half-way stage between using a high-level language and preparing the final machine code programs. Every type of microprocessor has its own set of symbolic instructions, in many cases similar to abbreviated English, which may be used to write programs. Programs written in symbolic code are known as **assembly code programs**. They must, of course, be first translated into machine code for direct use in the computer. A program which prepares the machine code program from a symbolic language program is known as an **assembler**.

1.2   PASCAL: origin and merits

PASCAL is named after the famous French mathematician Blaise Pascal, 1623–62, who invented one of the first practical calculators—the Pascaline, to help his father in tax calculations. It was developed by Professor Wirth of Zurich in 1970. He originally intended it to be used to help teach students to develop well-organised, structured programs, but since then PASCAL has gained very rapid popularity and is now generally regarded as one of the best general-purpose high-level languages for programming computers. PASCAL is ideally suited for engineering and scientific analysis and design work, and coupled with its availability for virtually all makes and types of computer—micros, minis and mainframes—it is now being widely adopted both in industry and education.

PASCAL enables problems to be tackled and solved using a "structured" approach by breaking the tasks into a number of logical steps. In the program solution, each step will be represented by writing a "procedure", each of which may be separately developed, tested and improved before being combined to form the complete program solution. Such an approach has a number of important merits: it forces you to think logically and systematically in constructing program solutions; it allows you to break up the problem into a number of simpler and more manageable steps; it allows you to develop and test each step individually; your resulting programs should be much easier to understand (by you after a lapse of time and, equally important, by another user) and much easier to check for errors.

1.3   An example of a program written in PASCAL

To provide an immediate idea of the form and the structure of a program written in PASCAL let us straightaway consider a practical example. Obviously at this early stage only a brief understanding will be gained—but, at least, it shows the general make-up of a PASCAL program and what is to be expected in later work.

Fig. 1.2 Series resonant R–L–C circuit

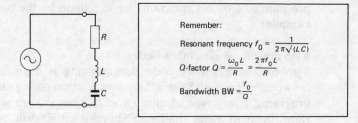

Remember:

Resonant frequency $f_0 = \dfrac{1}{2\pi\sqrt{(LC)}}$

Q-factor $Q = \dfrac{\omega_0 L}{R} = \dfrac{2\pi f_0 L}{R}$

Bandwidth $BW = \dfrac{f_0}{Q}$

The following program finds the resonant frequency, the Q-factor and half-power bandwidth for the simple resonant R–L–C circuit shown in fig. 1.2.

```
1   PROGRAM SERIESRES(input,output);

2   {* PROGRAM CALCULATES RESONANT FREQENCY,
3       Q-FACTOR AND 3 dB BANDWIDTH OF A
4       R-L-C SERIS CIRCUIT *}

5   CONST     PI=3.14159;

6   VAR       R,L,C,RESFREQ,Q,BW    :REAL;

7   PROCEDURE INPUTRLC;
8   BEGIN
9     WRITE('ENTER R VALUE IN OHMS   ');
10    READ(R);
11    WRITE('ENTER L VALUE IN HENRIES  ');
12    READ(L);
13    WRITE('ENTER C VALUE IN MICRO-FARADS  ');
14    READ(C)
15  END;

16  PROCEDURE CALCULATE;
17  BEGIN
18    RESFREQ:=1000/(2*PI*SQRT(L*C));
19    Q:=2*PI*L*RESFREQ/R;
20    BW:=RESFREQ/Q;
21  END;

22  PROCEDURE DISPLAYRESULTS;
23  BEGIN
24    WRITELN('***************************');
25    WRITELN('FOR R=',R,' L=',L,' C=',C);
26    WRITELN('RESONANT FREQUENCY=',RESFREQ:10);
27    WRITELN('      Q-FACTOR=',Q:10);
28    WRITELN('    3 dB BANDWIDTH=',BW:10);
29    WRITELN('***************************')
30  END;

31  BEGIN    { MAIN PROGRAM }
32    INPUTRLC;
33    CALCULATE;
34    DISPLAYRESULTS
35  END.
```

This program provides the values of the resonant frequency f_0, the circuit Q-factor, and the bandwidth for the values of the circuit components R, L and C entered from the computer keyboard. For example, when we execute the program, we obtain the following display on the VDU (the visual display unit); in this case, $10\,\Omega$ (for R), $0.005\,\text{H}$ (for L) and $0.1\,\mu\text{F}$ (for C) have been entered:

```
ENTER R VALUE IN OHMS   10
ENTER L VALUE IN HENRIES   0.005
ENTER C VALUE IN MICRO-FARADS   0.1
****************************
FOR R= 1.00000E+01 L= 5.00000E-03 C= 1.00000E-01
RESONANT FREQUENCY= 7.118E+03
        Q-FACTOR= 2.236E+01
    3 dB BANDWIDTH= 3.183E+02
************************
```

We now explain the significance and meaning of the various steps in the program, using for reference the line numbers given at the left-hand side of the program statements.

Note: these numbers would not be typed in when preparing the program; they may, however, be given automatically by your editor or in a listing when you "ask" the computer for one.

Line 1 All PASCAL programs start with the word PROGRAM followed by the program title. The title can be made up of only alphabetic characters, or a mix of alphabetic and number characters, but it *must* start with an alphabetic character.

The (input, output) following the program title is included to indicate to the compiler that the program is to have an input (the R; L and C data to be input via the keyboard) and is to produce an output (the values calculated for resonant frequency Q and bandwidth to be output to the screen or a printer).

Note: in many of the more recent implementations of PASCAL the "(input, output)" is omitted from the program heading and we need only type in (see also sections 2.8 and 6.5):

PROGRAM program-name;

The semi-colon ; is used in PASCAL to separate successive program statements, except the last statement which, as we shall see, is terminated by END.

Lines 2, 3, 4 Any characters included within the { } brackets are regarded as a COMMENT and will be skipped over in compilation (i.e. omitted when the PASCAL program is translated from the PASCAL source language to machine code by the compiler).

Comments are extremely useful for explaining what is being done by the program and/or by various sections of the program. Here we use the comment to define to the user what the whole program does. No semicolon should be used after the final comment } bracket.

Line 5 Following the program title and comment will be a declaration of any CONSTANTS to be used in the program. In our program, the constant $\pi = 3 \cdot 14159$ is declared by the statement:

CONST   PI = 3·14159;

Line 6 Following any constant statements, there must be a declaration of *all* the variables and their type that will be employed in the program solution.

This is done in the VAR declaration section. In this section, variables of the same type are given names (identifiers), each name being separated by a comma and the last variable in the list being followed by a colon. After the colon the type is specified.

In the present program variables have been used which only take "real" number values. This type is denoted in PASCAL by REAL. REAL variables correspond to decimal number values, e.g. 10·2, 74·83, 12·0, −9·89, as distinct from variables which take only whole number, INTEGER, values.

Thus, component values (R, L, C) and the values we wish to calculate (resonant frequency, Q, BW) must all be declared in the VAR section as follows:

VAR   R,L,C,RESFREQ,Q,BW :REAL;

This statement "tells" the computer to set aside storage locations within its RAM memory which can be used to store values for R, L, C, RESFREQ, Q and BW.

Lines 7–15, lines 16–21, lines 22–30 Following the PROGRAM heading, the CONST and VAR sections come any PROCEDURES or FUNCTIONS. (FUNCTIONS are not included in the present example.)

Each procedure can be considered as a routine or mini-program which will perform a set of prescribed tasks in the main program.

PROCEDURE INPUTRLC;

is used to effect the input of the R, L and C values via the keyboard.

Line 7 is used to define the procedure heading, i.e.

PROCEDURE name-of-procedure;

Line 8 BEGIN in PASCAL is always used to "begin" a group of executable statements—in this case, statements 9 to 14.

END; is used to terminate the group.

BEGIN and END; essentially act as brackets, to bracket an associated group of statements together.

Line 9 WRITE ('ENTER R VALUE IN OHMS   ');

is an example of PASCAL's basic output statement. In this case the computer is instructed by the WRITE statement to display the **string** (i.e. characters, words) between the single quotation marks. The whole of the string must be enclosed by the ( ) brackets and the statement separated from the next by ; as always.

Line 10 READ(R);

is an example of PASCAL's basic input statement. The READ statement "asks" us to input data from the keyboard. When this program is run, the display at line 9 appears, i.e.

ENTER R VALUE IN OHMS

and the cursor will flash, indicating that data input is required. Type in the R value, followed by a **carriage return** (or **enter** for some computers). The program execution then continues.

Lines 11 and 12 The WRITE statement displays:

ENTER L VALUE IN HENRIES

The READ(L) statement "asks" for an input. Input the L value.

Lines 13 and 14 The third pair of WRITE and READ statements is for input of the C value.

Line 15 END; is used to terminate the procedure.

Note: no semicolon is required in the penultimate line (line 14); the semicolon following END is sufficient.

BEGIN

⋮

executable statement(s)

⋮

END;

The BEGIN···END; construction is always used to bracket a number of statements together.

Lines 16–21 PROCEDURE CALCULATE does the actual calculation using the input data R, L and C to calculate RESFREQ (resonant frequency), Q and BW.

Line 18 is an example of an assignment statement and reads: "let the variable RESFREQ be assigned the value given by the right-hand expression." That is

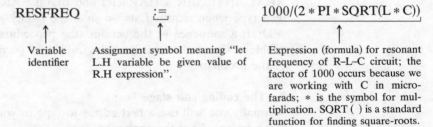

RESFREQ	:=	$1000/(2 * PI * SQRT(L * C))$
Variable identifier	Assignment symbol meaning "let L.H variable be given value of R.H expression".	Expression (formula) for resonant frequency of R–L–C circuit; the factor of 1000 occurs because we are working with C in micro-farads; ∗ is the symbol for multiplication. SQRT ( ) is a standard function for finding square-roots.

Line 19 is the assignment statement for Q:

Let Q be given the value $2\pi L f_0 / R$

Line 20 is the assignment statement for BW.

Lines 22–30 contain the PROCEDURE DISPLAYRESULTS and their function is simply to do that: the WRITELN statements (similar to WRITE but with a carriage return) display the results on the VDU.

Lines 31–35 form the **main program**:

BEGIN
INPUTRLC; ← calls procedure for inputting R, L, C values
CALCULATE; ← instructs computer to perform required calculations
DISPLAYRESULTS ← calls procedure to display results
END.

Note that all programs are terminated by END.

Don't forget the full-stop after END. This tells the computer at the compilation stage that this is the end of the program and no more lines are to be compiled.

1.4 An outline of the steps involved in writing and running PASCAL programs

To round off this introductory chapter we review the basic steps involved in writing and running a program. Fig. 1.3 provides a flowchart summary of the most important ones.

1 Problem definition

Understand exactly what you want of your program and clearly define your problem—obvious but absolutely essential as your first step. Note what information is given—this will constitute the data input to the program.

Define what information you require from the program solution—this will provide your output data to be obtained on running your program.

2 Source program development

Think out a suitable structure for your program. Note what constants can be usefully stated. List the variables and their type that need to be used and therefore declared in your program. (PASCAL has four standard types: REAL, INTEGER, CHARacter and BOOLEAN; you can also define your own type when required, as we shall see later.)

Draft a sequence of the actions (the procedures) required to solve your problem. Check that this leads logically to a correct and complete solution of the problem.

3 The coding/edit stage

Normally you will use a **text editor** to type in your program—at least your first version—from the keyboard.

An editor is a piece of software—essentially a wordprocessor—which allows direct entry of the program text from the computer keyboard with the ability to correct for mistakes in the entries, and to insert and delete lines, characters, etc. in the source program. When the source program has been entered in, it is normally saved on a floppy disc or magnetic tape so that it can be readily retrieved for processing whenever required.

Fig. 1.3 Flowchart summarizing steps in writing and running PASCAL programs

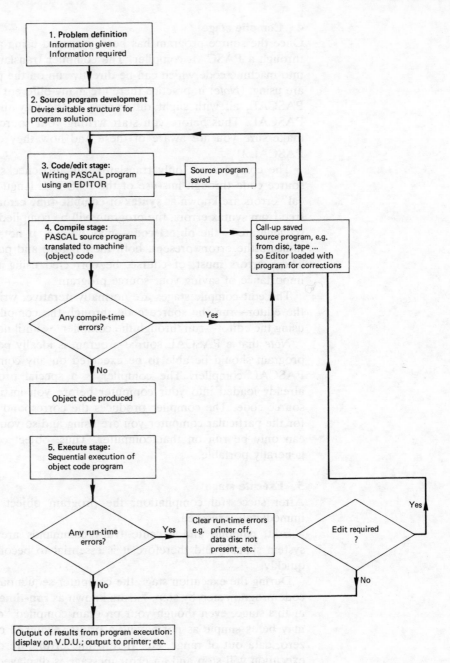

1. Problem definition
   Information given
   Information required

2. Source program development
   Devise suitable structure for program solution

3. Code/edit stage:
   Writing PASCAL program using an EDITOR

   → Source program saved

4. Compile stage:
   PASCAL source program translated to machine (object) code

   → Call-up saved source program, e.g. from disc, tape ... so Editor loaded with program for corrections

Any compile-time errors? — Yes →

No ↓

Object code produced

5. Execute stage:
   Sequential execution of object code program

Any run-time errors? — Yes → Clear run-time errors e.g. printer off, data disc not present, etc.

Edit required ? — Yes ↑

No ↓

No ↓

Output of results from program execution: display on V.D.U.; output to printer; etc.

4   Compile stage

Once the source program has been prepared using an editor, we "pass" it through a PASCAL compiler. The compiler translates the source program into machine code which can be directly run on the particular computer we are using. (*Note*: in practice there are many different compilers available for PASCAL, all with slight differences in the way they actually implement PASCAL. Thus before you start writing your source program you should make sure you are aware of these and how they differ from "standard" PASCAL.)

The compiler not only translates but also checks for any errors in the source code through mis-use of the PASCAL language. These "grammatical" errors are known as **syntax** or **compile-time errors**. If the source code is free from syntax errors, the program will be compiled and the machine code so generated—the **object code**—may be run. If not, your compiler should indicate the errors present, both their nature and position in the program. These errors must, of course, be corrected using the editor. Hence the importance of saving your source program.

The edit-compile stages are normally iterative: write your program using the editor—run the source code through the compiler—correct any errors using the editor—run through the compiler···until no errors are detected.

Note that a PASCAL source program is ideally portable. Any PASCAL program should be able to be executed on any computer supplied with a PASCAL compiler. The compiler is a special program which must be already loaded into your computer before you instruct it to translate the source code. The compiler produces the corresponding machine code only for the particular computer you are using and so your object code program can only be run on that computer. Thus, object code programs are not generally portable.

5   Execute stage

After successful compilation, the program object code is ready to be immediately executed.

Edit, compile and execute (run) commands are computer operating-system specific and therefore it is essential to become familiar with these quickly.

During the execution stage the computer sequentially executes the object code program step by step. Errors known as **run-time errors** may also occur at this stage, even though your program compiled "correctly". Such errors may be as simple as not having a printer switched on, trying to divide by zero, data out of range, etc. If run-time errors are detected then program execution will stop and an error message is displayed.

Obviously these errors must be debugged, either by clearing a peripheral fault or more usually by returning to the editor and repeating the edit/compile stages.

Even though no error is detected in executing your program, results should be very carefully scrutinized. Test, wherever possible, your program with input data producing known (and therefore checkable) results.

The fact that a program runs and produces results is little guarantee that your program might not still contain **logical errors**. There might still be mistakes in one or more of the algorithms (the systematic procedures you have employed to enable your problem to be solved in a finite number of steps).

2 PASCAL: Some Fundamentals

Introduction 13

2.1 Introduction and summary

In this chapter we consider some of the basics needed to begin writing
meaningful programs in PASCAL. Definitions are also given for some
important terms used in PASCAL: constants, variables, values and expressions. We will cover how variables are declared, how values are assigned,
how simple and compound statements are written, how information may be
input and output from the computer, and how data may be formatted (e.g.
expressed to a given number of decimal places).

2.2 Identifiers: syntax, standard and reserved words

The program title, all variables, constants, any procedures, functions, and
arrays that we use in a program are represented in PASCAL by identifiers.
Identifiers are the names we select to "identify" the variables, etc. used in
our program.

All identifiers in PASCAL must begin with one of the twenty six alphabetic letters. This letter can then be followed by any sequence of letters and/or
digits (i.e. 0, 1, 2, 3, · · · 9) but no other symbols.

Thus the following would be legal PASCAL identifiers:

```
NUMBER
VOLTAGE
TIMEDELAY
R11
CAP2
INDUCTORL3
```

whilst these would not be allowed, since they would lead to compile-time
errors:

1RESISTOR	an identifier must always start with a letter
V1+V2	+ symbol used; identifiers can only contain letters and digits
MIN VALUE	a space counts as a character in an identifier and therefore spaces cannot be used

Table 2.1 List of PASCAL reserved words

AND	END	NIL	SET
ARRAY	FILE	NOT	THEN
BEGIN	FOR	OF	TO
CASE	FUNCTION	OR	TYPE
CONST	GOTO	PACKED	UNTIL
DIV	IF	PROCEDURE	VAR
DO	IN	PROGRAM	WHILE
DOWNTO	LABEL	RECORD	WITH
ELSE	MOD	REPEAT	

Table 2.2 List of main PASCAL standard identifiers

Constants	Data types	Standard procedures		Standard functions		
FALSE	BOOLEAN	WRITE	NEW	SIN	SQRT	ORD
TRUE	CHAR	WRITELN	GET	COS	ABS	PRED
MAXINT	INTEGER	READ	PACK	ARCTAN	TRUNC	SUCC
	REAL	READLN	PAGE	EXP	ROUND	EOF
	TEXT	RESET	PUT	LN	ODD	EOLN
		REWRITE	UNPACK	SQR	CHR	

They are also a number of **reserved words** used in PASCAL statements, etc. and these obviously must not be used as identifiers. A list of the main reserved words used in PASCAL is given in table 2.1.

Likewise, PASCAL has a number of **standard identifiers**—the main ones are listed in table 2.2—which can only be used for their predefined purpose. These include standard PASCAL procedures, functions, operations, data types, etc.

An identifier may normally be of any length, although some implementations of PASCAL limit the recognisable length to a total of eight characters, i.e. the computer only takes into account the first eight letters and/or digits.

Always try to choose meaningful names for identifiers. This adds greater clarity to reading a program; it helps in understanding what the program is about, and is especially useful for debugging and future reference and invaluable for helping other users.

2.3 Syntax diagrams

Throughout the text we will be using diagrams known as **syntax diagrams** to help you follow the correct "grammatical" rules of PASCAL. When in doubt look up the appropriate syntax diagram.

Fig. 2.1 Syntax diagram for an **identifier**

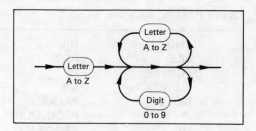

The syntax diagram for an identifier is shown in fig. 2.1. "Enter" the diagram at the left-hand side, always passing through a letter first, and then follow the arrow directions through any path you want. Provided you always move in the forward arrow directions you can pass around any loop as many times as you like.

Example 2.1 PASCAL make a distinction between whole number values (INTEGERs) and numbers which may or may not have a non-zero decimal part, known as REAL numbers. For example, 99·6, 70·0, −2·34 are examples of REAL numbers; 10, 12, −999 are examples of INTEGERs.

Fig. 2.2 Syntax diagrams for **INTEGER and REAL numbers**

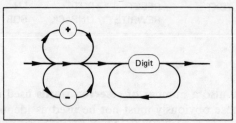

(a) INTEGER (whole) number

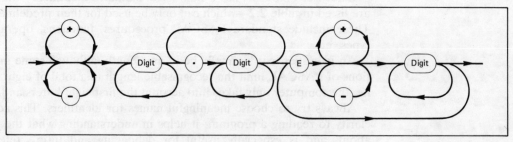

(b) REAL (decimal) number

Figure 2.2 shows the syntax diagrams for INTEGER and REAL numbers. Using these diagrams check whether

a) 55, −50, 72·0, +100, 6E3

have the correct syntax for INTEGER.

b) 62·0, −99·8, +10·00, 12E−4, ·0246

have the correct syntax for REAL numbers.

Solution

a) 55, −50, +100 are INTEGERs.

72·0 is not, it has a decimal point and zero.

6E3 which means 6×10^3 is incorrect; if it is to be written as an INTEGER, write as 6000.

b) 62·0, −99·8, +10·00, 12E−4 (=12×10^{-4}) are correct, but ·0246 is incorrect syntax; it should be written as 0·0246.

2.4 Constants, variables and their declaration

In most problems we are usually concerned with two basic kinds of data: **constants** and **variables**. PASCAL provides for both.

The value of a constant, as its name implies, remains constant throughout the execution of a program; whereas a variable may (although not always) have different values assigned to it as program execution proceeds.

Identifiers for all constants used in a program are declared (i.e. defined) immediately after the program heading, as illustrated in the following example:

```
PROGRAM   EXAMPLE1;
CONST     PI=3·1459;
          C=5E−6;
          R=1000;
          TOTALCOUNT=50;
          Y='YES';
          LINE='----------';
```

The reserved word **CONST** is used to instruct the computer at the compilation stage that storage locations for the following list of constant values are required. The constant identifiers must, of course, satisfy the correct PASCAL syntax. Values are assigned to each constant identifier using the equals symbol followed by the actual value. A semicolon is used to terminate each value definition.

Note: we can also define a constant as a letter, symbol, or string (string denotes any sequence of characters). In these cases the string values must be enclosed within the single quotation marks, as shown above.

Variables are declared (in the absence of any procedures and functions) immediately after the CONST section, using the reserved word **VAR**. In the VAR section we must specify both the *identifier* and its *type* for all the variables to be used in the program. VAR instructs the computer to provide storage locations for the values assigned to each of the variables declared. Once a value is assigned to a given variable it is stored in its location but, unlike a constant, the value may be changed as many times as directed by the execution of subsequent program statements.

Before we look at the form the VAR section takes, we shall briefly consider the concept of **variable type**. In PASCAL four standard types are used:

type	Examples of values taken by variable of this type
INTEGER	whole numbers, e.g. 0, 10, −99, 4167
REAL	decimal numbers, e.g. 10·0, −0·67, 5E−6
CHAR	single characters, e.g. 'A' · · · 'Z',
	'0', '1' · · · '9', '+', '*', '?'
	Note: character values are enclosed
	within the single quotation marks '.
BOOLEAN	Boolean variables can take only one
	of two values: TRUE or FALSE.

Declaration of all variables begins with VAR, followed by a list of identifiers (each separated by a comma), followed by a colon and their type specification. For example:

VAR   NOOFCOMPONENTS :INTEGER;
         R1,R2,L1,L2,C1 :REAL;
         COLOURCODE :CHAR;
         CIRCUITON :BOOLEAN;

One of the reasons for including the type specification is to aid the compiler in making error checks for mismatch in assigning a given type with a value inconsistent with that type.

The syntax diagram describing the form of the VAR declaration section is given in fig. 2.3.

Fig. 2.3 Syntax diagram for the **VAR declaration**

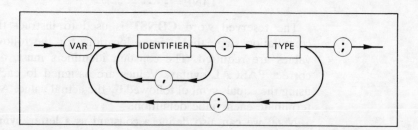

Remember

a) The VAR declaration section follows the program heading and the CONST declaration section.

b) It begins with VAR.

c) Variables of the same type are separated by a comma.

d) After the last variable of a given type write in a colon followed by the type, e.g. REAL, INTEGER, CHAR or BOOLEAN, and then (as always) a semicolon.

e) Note that spaces (except in identifiers) and a carriage return have no effect in the compilation of a program, so you can aid the layout of your program by using them.

Fig. 2.4

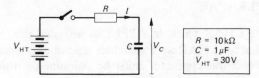

$$R = 10\,k\Omega$$
$$C = 1\,\mu F$$
$$V_{HT} = 30\,V$$

Example 2.2 Write a program to analyse the charge of the $C-R$ circuit shown in fig. 2.4, given the value of the supply voltage, R and C, which can then be declared as constants. The variables in the circuit analysis are current I, capacitor voltage V_C and time T.

The front-end of the program might then be

```
PROGRAM   RCTRANSIENT;
CONST      VHT=30;R=10E3;C=1E−6;
VAR        I,VC,T  :REAL;
BEGIN
. . . .
```

2.5. Assignment statements

Values are assigned to the variables we have previously declared in the VAR section by means of simple line statements known, not unexpectedly, as **assignment statements**. The general form of an assignment statement is shown in the syntax diagram of fig. 2.5.

Fig. 2.5 Syntax diagram for an **assignment statement**

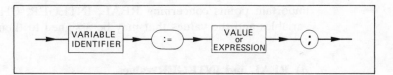

The symbol := is used in PASCAL to denote "is assigned the value of". Note that the equals symbol = must never be used as an alternative to := (colon key followed by equals key, when typing in) in an assignment statement.

Some assignment statement examples:

X:=5·2;

means the REAL variable X is assigned the value 5·2.

DAYOFWEEK:=3;

means the INTEGER variable DAYOFWEEK is assigned the value 3.

VOLTAGE:=10·2∗4·5;

means the REAL variable VOLTAGE is assigned the value of the expression 10·2∗4·5. (∗ is the multiplication symbol in PASCAL so VOLTAGE is assigned the value $10·2 \times 4·5 = 45·9$).

APPLE:='A';

means the CHAR variable APPLE is assigned the "value" A.

Note that for CHARacter variables we can only assign a single character as its value and this character must be contained within the single quotation marks.

HITEMP:=TEMP>=100·0;

means that the BOOLEAN variable HITEMP is assigned the value TRUE if the value of the TEMP variable is greater than or equal to 100·0 and FALSE if TEMP is less than 100·0.

2.6 Values and expressions: some introductory facts

In an assignment statement a value or, more generally, the value of an expression is assigned to a variable.

We will soon be meeting examples of expressions:

REAL expressions, e.g. $X-Y$, $X*Y$, $(25·7-5·63)/91·4$
INTEGER expressions, e.g. A DIV B, C MOD D
BOOLEAN expressions, e.g. P AND Q, NOT (P OR Q)

but at the moment it is enough to regard an expression as a "calculation/series of operations" that yields a value.

Before writing our first complete program it is necessary to clarify some important points concerning REAL, INTEGER, CHAR and BOOLEAN variables and the values that may be assigned to them.

a) **REAL and INTEGER values**

PASCAL distinguishes between INTEGERs (+ or − whole numbers) and REALs (numbers with a decimal part, even if the decimal part is zero).

There is, however, a practical limit to the maximum size of INTEGER that can be held in a computer, since computers have a finite word length, typically of 8, 16, 32 or 64 bits. For example, in a microcomputer with a 16-bit word length, the maximum INTEGER value is 32767. PASCAL includes a predefined identifier MAXINT which gives the maximum IN-TEGER value that can be held in a computer and so the range of INTEGER values for a given computer is

$$-MAXINT \cdots -3, -2, -1, 0, 1, 2, 3 \cdots +MAXINT$$

If this range is exceeded in program execution, the program execution stops and you will be notified that an **overflow error** has occurred.

REAL values are represented in a computer differently from INTEGERs to allow for a much greater range of values. There are two ways of writing REALs in PASCAL programs:

Fixed point notation (normal decimal)

e.g. 10·0, 74·92, −174·0, 0·31478, −0·00781

Floating point notation

e.g.   1·26E3 which means $1·26 \times 10^3 = 1260·0$
         4E−2 which means $4 \times 10^{-2} = 0·04$
    299·7E+6 which means $299·7 \times 10^6$

It is important to remember that the representation of REAL numbers in a computer is not exact, since computers store REALs in a finite word length. The accuracy limit varies from computer to computer but is normally at least to 8 significant figures and can be as much as 18. Although this may not be of much worry in the majority of problems, such errors can accumulate and give erroneous results, especially in iterative routines employed in many numerical method solutions. Remember this and guard against testing almost equal numbers for equality and subtracting almost equal numbers.

b) **CHARacter values**

The characters, i.e. letters, digits, punctuation marks, space, +, −, etc., that appear on the keyboard together with the control characters constitute the *character set* of the computer. In PASCAL a variable of the data type CHAR is defined as a character within the available character set of the computer.

Thus a CHAR variable may be assigned any character available in the set, which normally includes:

(i) Upper and lower case letters

    A, B, C, D · · · X, Y, Z   a, b, c, d · · · x, y, z

(ii) The ten denary digits

    0, 1, 2, 3, 4, 5, 6, 7, 8, 9

(iii) Punctuation marks

    ! , . ; ? : ” ’

(iv) Arithmetic operators

    + − * / ( )

(v) Space and graphical symbols

The character value must always be entered in between the single quotation marks in an assignment statement, e.g.

```
SPACE:=' ';
ADDSIGN:='+';
LITTLEA:='a';
```

c) **BOOLEAN values**
Boolean variables can only be assigned one of two values: either TRUE or FALSE.

Boolean values very often arise from the result of some comparison, e.g.

LIMIT:=CURRENT>=10·0;

assigns the BOOLEAN variable LIMIT the value TRUE if the (REAL) variable CURRENT has the value equal to or greater than 10·0, but the value FALSE if the value of CURRENT is less than 10.

2.7 Obtaining an output from a program: WRITE and WRITELN statements

The **WRITE** statement is the basic output statement in PASCAL. It is used to instruct the computer to make an output display on the VDU (visual display unit) and/or a printer (or other output device) if connected and switched on.

WRITE statements may be used to effect calculations and display their result and to display strings. For example:

1 WRITE(calculation to be performed);

so   WRITE(25+75);

would produce the result of 25+75, i.e. 100 would be displayed.

X:=4·6;
Y:=12·2;
WRITE(Y−X);

would produce the result of 12·2−4·6, i.e. 7·6 would be displayed.

2 WRITE('any string of characters');
This statement would cause the display of the character string enclosed within the single quotation marks to be displayed.

3 Combination of string and data output in one WRITE statement:

WRITE('··· string ···',data value/calculation);

display the string and data value, e.g.

WRITE('RESISTANCE=',25·6/4);

would cause

RESISTANCE=6.4

to be displayed.

The **WRITELN** statement performs the same output action as WRITE but in addition causes a carriage return to be made *after* the WRITELN statement has been obeyed. Any subsequent output will then be displayed on the next line. For example,

WRITELN('THE VALUES OF VOLTAGE AND CURRENT ARE:');
WRITELN('V=',9·6*0·15);
WRITELN('I=',7·2/28·8);
WRITELN('- - - - - - - - - - - -');

would produce the output display:

THE VALUES OF VOLTAGE AND CURRENT ARE:
V = 1·44
I = 0·25
- - - - - - - - - - - -

Note that the data values 1·44 and 0·25 may be output in floating point form:

1·44000E+00 and 2·50000E−01

for example. In either case, when REAL or INTEGER values are output they are normally right-aligned, which is a useful form when tabular representation of data is required.

The basic form of WRITE and WRITELN statements is summarized in the syntax diagram of fig. 2.6.

One final point, useful for producing a line feed or a blank line in a display, is the statement

WRITELN;

which when used alone—no brackets or quotes—enforces a carriage return.

Fig. 2.6 Syntax diagram for simple **WRITE and WRITELN** statements

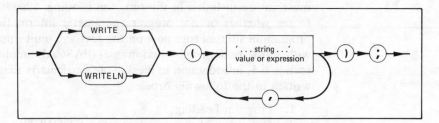

2.8 Comments: use to aid program understanding

Comments may be made anywhere within a program and are extremely useful to explain the general purpose of the program and its various sub-sections.

Comments are placed within curly brackets { } or between (* *), as indicated in the syntax diagram of fig. 2.7, and may be placed anywhere after a space or at the end of a line in the program. All comments are

Fig. 2.7 Syntax diagram for **comment**

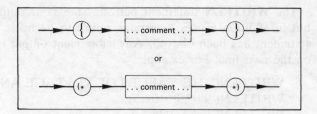

skipped over at compilation stage and in addition to their proper purpose of aiding program explanation, the comment-type brackets can be used to help error diagnosis. For example, if it is expected that part of a program contains an error, it may be bracketed off and then, if the rest of the program compiles, we then know that a possible error exists in the section isolated within the comment brackets.

2.9 Beginning to write simple programs

We are now in a position to write simple programs: we have considered how to define constants and variables, how to assign values to variables, and how to use the WRITE statements to obtain an output display of our results.

The structure and order of a simple program is shown in fig. 2.8. The inclusion of (input, output) in the program heading in fig. 2.8 specifies that the program is to perform both input and output operations. Input and output are called **program parameters**. Input is included when data is to be **read into** the program; output is included when the program is to **write out** data. However, in many implementations of PASCAL, program parameters are not specified in the program heading. The read and write statements in the program automatically set up lines of communication to the input keyboard and the output VDU or printer. In these cases, (input, output) must not be included in the program heading. Check your implementation to see whether or not program parameter information is required in the program heading. From now on the text will omit (input, output) consistent with most modern (non-mainframe) PASCAL implementations (see also section 6.5, introduction to files). Note particularly that the program must be written in the following order:

1. Program heading;
2. Definition of any constants (using CONST);
3. Declaration of all variables to be used and their type (the VAR section);
4. BEGIN
   ⋮
   program statements, e.g. assignment, WRITE, etc.;
   ⋮
   END.

Note also that the executable statements in the program always begin with BEGIN and the final terminating line of the program is always END.

Fig. 2.8 Diagram showing structure and order of a simple program [*Note:* the (input, output) program parameters in the program heading are not included in most of the more recent implementations of PASCAL but are usually required in mainframe systems.]

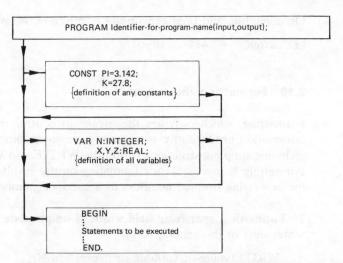

BEGIN · · · and · · · END; are also used to bracket together a number of statements making up a compound statement (see section 2.12). Semicolons are used to separate successive statements, except the ones preceding END, so don't forget these!

Example 2.3 Write a program to find the resistance of a 100 m copper wire of circular cross-section, radius 1 mm. The resistivity of copper $\rho = 1.56 \times 10^{-8}$ Ωm and the formula relating resistance to its resistivity, length, cross-section area is

$$R = \rho \times \text{length}/\text{area}$$

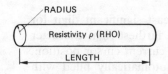

Fig. 2.9 Resistance =

$$\text{RHO} \times \frac{\text{LENGTH}}{\text{AREA}}$$

Solution A typical program could be of the following form where we define two constants π and resistivity, i.e. PI and RHO, and declare three variables: LENGTH, RADIUS, RESISTANCE of type REAL. The executable part of the program consists of three assignment statements, the third applying the formula $R = \rho l / A$, and one WRITE statement to output the result.

```
PROGRAM RESISTCALC;

{ PROGRAM CALCULATES RESISTANCE OF COPPER WIRE
  OF LENGTH 100 m , RADIUS 1 mm }


CONST  PI=3.14159;
       RHO=1.56E-08;

VAR    LENGTH,RADIUS,RESISTANCE :REAL;

BEGIN
  LENGTH:= 100.0;
  RADIUS:= 0.001;
  RESISTANCE:= RHO*LENGTH/(PI*RADIUS*RADIUS);
  WRITE('RESISTANCE = ',RESISTANCE)
END.
```

On compiling and then executing the program, we obtain the output:

```
RESISTANCE =   4.96564E-01
```

2.10   Formatting the output

Formatting, which provides the means to control the form of the output information produced by a program, is easily accomplished in PASCAL by including simple instructions within the WRITE and WRITELN statements. Formatting is invaluable for tabulating output results and especially useful for expressing decimal numbers to a specified number of decimal places.

1   Formatting: specifying field width of output data
Statements of the general form,

> WRITE(value-of-variable or expression:N);
> WRITELN(VALUE1:10);
> WRITE(VALUE1:N1,VALUE2:N2,VALUE3:N3);

> where N, N1, N2, N3 are integer values,

instruct the computer to display or print out the value(s) of the variable(s) in a given number of spaces; this number, known as the **field width**, is the integer value (N, 10, N1, N2, N3, etc.) following the colon in the WRITE and WRITELN statements.

The data value is printed or displayed so the least significant digit for INTEGERs, the character for CHAR values or the farthest right character for strings is displayed N spaces to the right of the current cursor position. Any "unused" spaces in the field width N are automatically filled with blanks.

Running the following program illustrates the form of display obtained, when the WRITELN statements contain the field-width formatting information described above.

Example 2.4

```
PROGRAM T2;

CONST A='A'; B='B';
      STRING='*****12345';

VAR  C,D       :CHAR;
     N1,N2,N3  :INTEGER;
BEGIN
 C:='C';D:='D';
 N1:=23;N2:=3478;N3:=993;
 WRITELN(A:10,B:10);
 WRITELN(STRING:20);
 WRITELN(C:10,D:10);
 WRITELN(N1:8,N2:8,N3:8)
END.
```

On executing this program, the following display is obtained:

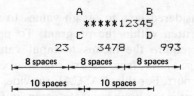

2 Formatting REALs: specifying decimal numbers to a given number of places

REAL values may be output to a given number of decimal places using WRITE statements of the form:

WRITE(value-of-real-variable-or-expression:N:P);
WRITE(VALUE1:N1:P1,VALUE2:N2:P2);
WRITELN(TEMP:5:2,BETA:4:1,R:6:3);

The first integer after the *first* colon defines the field width. The integer after the *second* colon specifies the number of decimal places to be given in the display. Example 2.4 illustrates the use of

WRITE(VALUE:N:P)

where N = total field width, P = number of decimal places.

Example 2.5

```
PROGRAM T3;

VAR  X,Y    :REAL;

BEGIN
X:=34.872;Y:=-6.683;
WRITE(X:10:2,Y:10:2);
WRITELN(X+Y:10:1,X-Y:10:2)
END.
```

On running this program, we obtain the following display:

```
        34.87       -6.68        28.2       41.55
```
|←10 spaces→|←10 spaces→|←10 spaces→|←10 spaces→|

2 decimal places   2 decimal places   1 decimal place   2 decimal places

2.11 Inputting data to a program: READ and READLN statements

So far we have considered how to assign values to variables using assignment statements written within the program. To provide true interaction with programs we require the means to input values, data, etc. via the keyboard or, as we see later, from an external source such as from a file on disc or from an I/O port. For this, PASCAL provides the **READ** statement. The READ statement is the basic input statement in PASCAL and takes the general form (see also fig. 2.10):

READ(VARIABLE   IDENTIFIER);

or if more than one variable value is to be read in:

READ(IDENTIFIER1,INDENTIFIER2,IDENTIFIER3);

Fig. 2.10 Syntax diagram for **READ** and **READLN,** the basic input statements

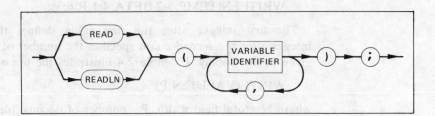

When a READ statement is met in program execution, a line of communication is opened up to the keyboard. Program execution stops and awaits your entry from the keyboard. The "await" state is usually indicated by the flashing of the cursor or by a ?. Make your entry followed by a carriage return (or pressing the ENTER key if your computer keyboard has this rather than carriage return). This action provides the program with the data value and execution immediately continues.

For a "multi-value" read statement of the form:

READ(X,Y,Z);

where X, Y, Z are either REALs OR INTEGERs, the corresponding value for each must be entered in the same order as appears in the READ statements, and each value must be separated by a space. After entering the last value press carriage return (or ENTER) for execution to continue.

The application of the READ statement is illustrated in the following example.

Example 2.6 The following program calculates the total resistance of any three resistors connected in parallel.

Fig. 2.11 Total resistance *R* of parallel circuit

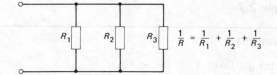

$$\frac{1}{R} = \frac{1}{R_1} + \frac{1}{R_2} + \frac{1}{R_3}$$

```
PROGRAM T4;

{ FINDS RESISTANCE OF 3 RESISTORS IN PARALLEL }

VAR  R,R1,R2,R3  :REAL;

BEGIN
 WRITE('ENTER VALUES OF THE 3 RESISTORS  ');
 READ(R1,R2,R3);
 R:=1/(1/R1+1/R2+1/R3);
 WRITELN('TOTAL RESISTANCE = ',R:10:2)
END.
```

On executing the program, the string in the WRITE statement is first displayed. We then enter the values for R1, R2 and R3:

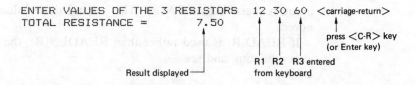

The READ statement can be used to input REAL, INTEGER and CHAR values but not BOOLEAN or other user-declared types. Each variable in the READ statement must be matched with the correct type of data read in, e.g. a REAL value must correspond to a REAL variable in the READ statement. If not, a mismatch error will be registered and program execution will halt.

For some applications it may be required for the input data to be read in one line at a time. For these cases, PASCAL provides the **READLN** statement, e.g.

READLN(X,Y,Z);

READLN is similar to READ but, after the READLN statement has been obeyed, any unread data in the current line of input (e.g. from a disc file) is skipped and any subsequent input request is taken from the next line of data. The action of READLN is illustrated in the following example.

Example 2.7

```
PROGRAM T5;

{ TO ILLUSTRATE USE OF READLN }

VAR R,C:INTEGER;

BEGIN
 WRITE('ENTER NO. OF RESISTORS ');
 READLN(R);
 WRITE('ENTER NO. OF CAPACITORS ');
 READ(C);
 WRITELN('NO. OF RESISTORS = ',R);
 WRITELN('NO. OF CAPACITORS = ',C)
END.
```

On executing the program:

```
ENTER NO. OF RESISTORS 12236 RESISTORS  <C-R>

ENTER NO. OF CAPACITORS 2500 CAPACITORS <C-R>
NO. OF RESISTORS = 12236
NO. OF CAPACITORS = 2500        Entries made by us
                                <C-R> indicates carriage
                                return is pressed
```

You can see that the program outputs the correct results despite having entered 12236 RESISTORS—two items of data—in response to the first input request: READLN(R). The second data value RESISTORS is ignored.

If READ(R) is used rather than READLN(R), the program "fails"—try experimenting and see.

2.12 Simple and compound statements

In PASCAL there are basically two types of statement:

Simple statements: statements that cannot be grammatically divided; an assignment statement is an example of a simple statement.

Compound statements: statements which consist of a number of simple statements preceded by BEGIN and terminated by END; they are extremely useful wherever a group of instructions is required to be considered as one.

The executable sections of a PASCAL program normally consist of a combination of simple and compound statements.

The main objective of a compound statement is to cause a sequence of simple statements to be bracketed together and treated as a single statement for syntax purposes. This is done within the program by bracketing the required sequence of statements between BEGIN and END. It is good practice to indent a compound statement by starting each line making up the compound statement with two or more spaces, although normally not more than five.

The syntax diagram for a compound statement is shown in fig. 2.12.

Fig. 2.12 Syntax diagram for a **compound statement**

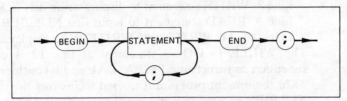

The following program, which may be used to find the average of any number of values entered via the keyboard, contains a compound statement (in this case a group of statements to read in and sum values) embedded in the main program.

Example 2.8

```
1    PROGRAM T6;

2    {PROGRAM FINDS AVERAGE OF DATA ENTERED
3     IN FROM KEYBOARD}

4    VAR      NUMBER,SUM        :REAL;
5             NOOFTERMS         :INTEGER;

6    BEGIN
7      NOOFTERMS:=0;SUM:=0.0;
8      WRITE('ENTER FIRST NUMBER ');
9      READ(NUMBER);
10     WRITE(' After entering last number,');
11     WRITELN(' type -9999.0 to finish');
12       WHILE NUMBER<>-9999.0 DO

13       { Beginning of compound statement }
14       BEGIN
15         NOOFTERMS:=NOOFTERMS+1;
16         SUM:=SUM+NUMBER;
17         WRITE('ENTER NEXT NUMBER: ');
18         READ(NUMBER)
19       END;
20       { End of compound statement }

21     WRITELN('AVERAGE = ',SUM/NOOFTERMS:8:2)
22 END.
```

Here is an example of running the above program:

```
ENTER FIRST NUMBER 12.3
 After entering last number, type -9999.0 to finish
ENTER NEXT NUMBER: 56.9
ENTER NEXT NUMBER: 45.7
ENTER NEXT NUMBER: 38.64
ENTER NEXT NUMBER: 21.0
ENTER NEXT NUMBER: -9999.0
AVERAGE =      34.91
```

The main program starting with BEGIN at line 9 and terminating with END. at line 22 contains a number of simple statements, e.g.

Line 7: assignment statements to initialize NOOFTERMS and SUM to zero.

Line 8: WRITE statement to display string 'ENTER FIRST NUMBER'.

Line 9: READ statement to input first NUMBER.

Lines 10, 11: WRITE and WRITELN to display string information.

The WHILE··· DO··· statement at line 12 is one of three forms of repetition constructs used in PASCAL and is considered in detail in Chapter 4. In the present program it is used to instruct the computer "WHILE the NUMBER entered in is not equal to −9999.0, DO the compound statement bracketed together by BEGIN and END; in lines 15 to 18".

The compound statement itself (apart from the comments which anyway are skipped over) is composed of a group of simple statements:

Line 15: increment NOOFTERMS (this acts as a count of number of values entered)

Line 16: sums values entered

Line 17: displays string 'ENTER NEXT NUMBER'

Line 18: requests input of next number

The compound statement will be continually obeyed until we enter −9999.0. When this is entered, the WHILE loop is exitted and execution proceeds to line 21 which displays the result.

Note that the final END. at line 22 is used to terminate the program. Don't forget the full stop after END.

2.13 TYPE: user-defined enumerated scalar and subrange data types*

a) Introduction

So far we have seen that PASCAL provides four standard data types for variables used in programs, i.e. INTEGER, REAL, CHAR and BOOLEAN.

INTEGER, CHAR and BOOLEAN are referred to as **ordinal** types since they represent an ordered set of values, i.e. the values occur in an ordered sequence, are countable and finite in number, and may be compared with one another. The total number of values in an ordinal data type is known as the **cardinality** of the data type, e.g. the cardinality for BOOLEAN data type is 2 (TRUE and FALSE values); the cardinality for INTEGERs is

$$2 \times MAXINT + 1 \qquad (-MAXINT \cdots 0 \cdots +MAXINT).$$

PASCAL also allows us to define two kinds of user-defined ordinal data types: the **enumerated type** and the **subrange type**, which are dealt with below.

---

* This section may be omitted on first reading.

User-defined types have a number of advantages, and a few disadvantages, at least for the more casual programmer. PASCAL is known as a **strongly-typed** language, "strong" not indicating that you have to hit the keyboard with great force but referring to the distinct types of data that can be used in programs and the fact that you cannot mix up types in PASCAL. For example, you cannot assign to a variable declared as INTEGER a CHAR data value: you will be informed of "type-mismatch" or "data-mismatch error" if you try to do so.

However, strong typing and the ability to define your own type can be a great asset. Many problems can be expressed more clearly by utilising user-defined type; reading and writing the program solution can be made easier; user-defined type provides additional information to the compiler and this information can be used to make a more extensive check for errors and also normally leads to the production of a more efficient code for programs.

User-defined data types are declared in the TYPE section, which follows the CONST and precedes the VAR declaration sections, i.e.

CONST   (definition of constants);
TYPE   (definition of types);
VAR   (definition of variables and their type);

b) Enumerated scalar type
Enumerated (meaning ordered and countable) scalar (meaning single-valued items) type is defined as follows:

TYPE   *type-identifier*=(*constant-1,constant-2,*···);

where **TYPE** is a reserved word used in PASCAL to precede user-defined type definitions; *type-identifier* is the identifier used to name the new data type and *constant-1*, *constant-2*, etc. are the constants (values) which the new data type may take.

For example,

TYPE   TIMEUNITS=(SEC,MIN,HOUR);

defines a new data type called TIMEUNITS, which can take the 3 values: SEC, MIN or HOUR and is ordered so that

SEC < MIN < HOUR

Variables of a user-defined enumerated type may then be declared in the VAR section in the usual way, e.g.

VAR   DELAY,PERIOD,TESTTIME   :TIMEUNITS;

i.e. the variables DELAY, PERIOD, TESTTIME are now defined as variable of type TIMEUNITS.

The TYPE-VAR declaration can be combined into a single declaration in the VAR section:

VAR   *variable-identifier-1,*···:(*constant-1,constant-2,*···);

For example,

TYPE   TIMEUNITS=(SEC,MIN,HOUR);
VAR   DELAY,PERIOD,TESTTIME   :TIMEUNITS;

can be combined and declared in shorthand form by

VAR   DELAY,PERIOD,TESTTIME   :(SEC,MIN,HOUR);

c) Some basics for enumerated scalar data types
1 A data value (constant) may not belong to more than one type. For example,

TYPE   TIMEUNITS=(SEC,MIN,HOUR,DAY);
         DATE   =(DAY,MONTH,YEAR);

both contain the value DAY. DAY cannot appear in both value lists. We can, of course, overcome the problem by using a different identifier in each list, e.g. change DAY to DAYS in the TIMEUNITS list.

2 Assignment and tests for equality can normally be applied to all data types. However, most operations provided in PASCAL are limited to the type of value to which they may be sensibly applied. For example, we can write

DELAY:=HOUR;
PERIOD:=MIN;
IF PERIOD>SEC THEN (*statement 1*)
ELSE (*statement 2*);

The first two of the above statements assign respectively the value of HOUR to the variable DELAY and MIN to the variable PERIOD. The IF··· THEN··· ELSE conditional statement (see Chapter 4, section 4.2) would cause *statement 1* to be executed if PERIOD were greater than SEC, i.e. either MIN or HOUR, otherwise *statement 2* would be executed.

4 It is not possible to READ or WRITE a user-defined enumerated scalar value directly, e.g.

PERIOD:=SEC;
WRITE(PERIOD);   ← not permitted
READ(PERIOD);   ← not permitted

Enumerated scalar type, however, may be used as the control variable in FOR statements, as selectors in CASE statements (see Chapter 4) and with certain standard functions (CHR, ORD, SUCC, PRED, see Chapter 3).

d) Subrange types
For any enumerated scalar type, user-defined or INTEGER, CHAR, BOOLEAN (but not REAL), it is possible to create a subrange type whose values are a subrange of the original type members.

A subrange type is defined as follows:

TYPE   *type-identifier*=*lower-limit* **. .** *upper limit*;

where *lower* and *upper-limit* are constants specifying the lower and upper values in the subrange. These constants are separated by two dots **. .** , which mean "through" the range of values bounded by and including the two constants specified. The lower-limit constant must have a lower ordinal value than the upper-limit constant and both constants must, obviously, be of the same type.

The use of subrange types has a number of real advantages. They add to the clarity of the program, they may save memory space, and they certainly provide a valuable aid to error checking by ascertaining that a variable value is within range.

For example,

(1)   TYPE   POSITIVENO=1 . .1000;
      VAR   N :POSITIVENO;

defines a subrange type POSITIVENO to include the positive numbers $1, 2, 3 \cdots$ to 1000; the variable N is then declared in the VAR section as a positive number. Thus, throughout the program N can only be assigned numbers within the range 1 to 1000; if through any operation N takes on a value outside its defined range, we will be notified of "an out-of-range error" and program execution will be halted.

(2)   TYPE   CAPLETTER = 'A' . .'Z';
      VAR   CHA :CAPLETTER;

restricts the variable CHA to capital letters of the alphabet.

(3)   TYPE   COLOURCODE=(BLACK,BROWN,RED,ORANGE,
                          YELLOW,GREEN,BLUE,VIOLET,
                          GREY,WHITE);
              COLOURØ123=BLACK . .ORANGE;
      VAR   BANDCOLOUR:COLOURCODE;
              LOWCOLOUR,BANDØTO3:COLOURØ123;

The enumerated type COLOURCODE is defined to consist of a set of 10 values: BLACK, BROWN, $\cdots$ WHITE. A variable of type COLOUR-CODE can be assigned any one of these values; thus BANDCOLOUR can take BLACK or BROWN or RED, $\cdots$ or WHITE as a value.

The subrange type COLOURØ123 restricts a variable of its type to one of the four values: BLACK, BROWN, RED, ORANGE; thus the variables LOWCOLOUR, BANDØTO3 are restricted to this subrange.

Shorthand notation can also be used to combine TYPE and VAR into a single declaration in the VAR section. For example,

TYPE   POSITIVENO=1:1000;
VAR   N :POSITIVENO;

can be written in a single declaration,

    VAR   N :1..1000;

READ and WRITE procedures, which cannot be used with the enumerated type, can be used with subrange types when the family type from which the subrange is formed consists of integers or alphabetic characters—ordinal type—known to the computer. For example,

    READ(N);WRITE(N);   {N:1..1000}
    READ(CHA);WRITE(N);   {CHA:'A'..'Z'}

can be used to input and output information, but

    READ(LOWCOLOUR);
    WRITE(LOWCOLOUR);{LOWCOLOUR:BLACK..ORANGE}

cannot.

Exercises 2

2.1 State which of the following are *not* valid PASCAL identifiers. Give also your reason

R	R1VALUE	CONST	noofvalues
1XRAY	X+Y	HI-TEMP	NO OF VALUES
X11	VAR	'ABC'	WRITE
ohms	+Z	INPUT	MAXMINLIMITS

2.2 With reference to PASCAL programs, distinguish between the following:
a) Constants and variables
b) Real and integer values
c) CHAR and BOOLEAN variables
d) READ and WRITE statements
e) Simple and compound statements
f) FALSE and TRUE
g) := and = symbols

2.3 Write a PASCAL program to find the average value of the following quantities which are to be entered from the keyboard:

12·671   15·236   4·007   5·569   14·798   21·003

Output the result accurate to one decimal place.

2.4 Write a PASCAL program to evaluate the resistance of two resistors R_1 and R_2 connected in parallel. The values of R_1 and R_2 are to be entered from the keyboard and the result is to be specified to the nearest whole number. Remember:

$$\frac{1}{R} = \frac{1}{R_1} + \frac{1}{R_2}$$

3 Expressions, Calculations and Standard Functions

3.1 Introduction and summary

The main purpose of this chapter is to provide a practical description of the operators, expressions and standard functions available in PASCAL and to use these in programs.

In particular we consider how calculations are performed in PASCAL using the basic REAL and INTEGER operators. We consider also the meaning of comparison and Boolean type operators and expressions and how they may be used to provide the "test" conditions in practical programs. Finally we consider the more important standard functions available in PASCAL for executing prescribed tasks, e.g. for finding square roots, trigonometric functions, exponential and log, and functions used in association with the CHARacter data.

3.2 Arithmetical calculations: REAL expressions and arithmetic operators

In the last chapter we used fairly freely the basic arithmetic operators to perform calculations. The four basic arithmetic operations $(+ - \times \div)$ are normally performed with REAL data:

for ADDITION   use the $\boxed{+}$ key, e.g. W: = A + B
for SUBTRACTION   use the $\boxed{-}$ key, e.g. X: = A − B
for MULTIPLICATION use the $\boxed{*}$ key, e.g. Y: = A ∗ B
for DIVISION   use the $\boxed{/}$ key, e.g. Z: = A/B

where W, X, Y, Z are REAL variables and A and B are REAL or INTEGER data values

The normal rules of precedence apply, i.e

∗ and / (multiplication and division) before + and −

Brackets ( ), not the curly { } or [ ], can be used to bracket terms in an expression in the usual way, e.g.

X: = (A − B)/(A + B)

meaning X is assigned the value $\dfrac{A - B}{A + B}$

$$Y:=23{\cdot}0*A*(6{\cdot}91-5{\cdot}84*B)$$

meaning $23{\cdot}0A(6{\cdot}91-5{\cdot}84B)$

Note that, in standard PASCAL, INTEGER values may be used in a REAL expression without qualification. INTEGER values are automatically converted to REALs in program execution. Note also that no expression is directly available in PASCAL for calculating powers, e.g. X^6, $X^{1{\cdot}4}$, $X^{-3{\cdot}2}$; we explain how powers may be evaluated in example 3.9 of this chapter and section 5.6 of Chapter 5.

Example 3.1 This program illustrates some basic calculations in PASCAL. Remember that WRITE or WRITELN is used to effect the calculation and output the result, and that the formatting information defines the field width and number of decimal places given in the result.

```
PROGRAM CALC1;

{ILLUSTRATION OF SOME BASIC CALCULATIONS}

CONST  PI=3.14159;

VAR    A,B,R :REAL;
       N,M   :INTEGER;
BEGIN
 A:=42.0; B:=15.5 ; R:=22.8;
 M:=12; N:=67;
  WRITELN('A+B = ',A+B:5:1);
  WRITELN('A-B = ',A-B:5:1);
  WRITELN('AxB = ',A*B:6:2);
  WRITELN('A/B = ',A/B:6:3);
  WRITELN('2xPIxR = ',2*PI*R:8:3);
  WRITELN('M/N = ',M/N:6:3);
  WRITELN('(A+B)/(A-B) = ',(A+B)/(A-B):10:5)
END.
```

On running the program, we obtain the following output display:

```
A+B =  57.5
A-B =  26.5
AxB = 651.00
A/B =  2.710
2xPIxR =  143.256
M/N =  0.179
(A+B)/(A-B) =    2.16981
```

Example 3.2 Write a program which outputs the value of the voltage V_D for a value of temperature entered in via the keyboard, for the bridge circuit of fig. 3.1 given:

$$V_S = 10 \text{ V}$$
$$R_1 = 470\ \Omega,\ R_2 = 330\ \Omega,\ R_3 = 580\ \Omega$$

Fig. 3.1 Bridge circuit

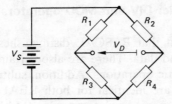

$$R_4 = 220\,(1+\alpha T),\ T = \text{temperature } ^{\circ}\text{C},\ \alpha = 0{\cdot}005\,^{\circ}\text{C}^{-1}$$

$$V_D = \left[\frac{R_2 R_3 - R_1 R_4}{(R_1 + R_3)(R_2 + R_4)}\right] VS$$

Check your program for the cases $T = 20$, $T = 100$, $T = -21{\cdot}7\,^{\circ}\text{C}$ where you should obtain $V_D = 1293$, $523{\cdot}8$, 1796 mV respectively.

```
PROGRAM BRIDGECALC;

{ TO FIND VOLTAGE VD FOR BRIDGE CIRCUIT }

CONST   VS=10; ALPHA=0.005;
        R1=470;R2=330;R3=580;

VAR     VD,T,R4:REAL;

BEGIN
 WRITE('ENTER VALUE OF TEMPERATURE, T= ');
 READ(T);
 R4:=220*(1+ALPHA*T);
 VD:=(R2*R3-R1*R4)*VS/((R1+R3)*(R2+R4));
 WRITE('FOR T=',T:6:1,' VD=',1000*VD:6:1,' mV')
END.
```

On running the program:

```
FOR T= -21.7 VD=1796.0 mV
ENTER VALUE OF TEMPERATURE, T= 20

FOR T=  20.0 VD=1293.0 mV
ENTER VALUE OF TEMPERATURE, T= 100

FOR T= 100.0 VD= 523.8 mV
ENTER VALUE OF TEMPERATURE, T= -21.7
```

The program BRIDGECALC is one possible solution. Here we have defined the given data as constants and declare the required voltage VD, temperature T and the temperature-sensitive resistance R4 as REAL variables.

In the main program we use a READ statement to input T, R4 is then assigned the value $220\,(1+\alpha T)$ and VD the expression given. The final WRITE statement outputs the result.

3.3 Integer arithmetic: DIV and MOD operators and expressions

We have already noted that PASCAL distinguishes between REAL and INTEGER variables and data. There are also certain important differences in dealing with arithmetic operations. Addition, subtraction and multiplication operators $(+ - *)$ are identical for both REAL and INTEGER variables and, provided the maximum integer value (MAXINT) is not exceeded, they can be used in the same way as REALs. Division, however, is different. The / operator is not normally used with INTEGER variables; if used it will return a REAL result.

PASCAL uses the operators **DIV** and **MOD** to perform "division" with INTEGER numbers. The DIV operation performs **division with truncation**, i.e. gives the integer part of the result only and discards any fractional part.

> X: = 25 DIV 7;

assigns the INTEGER variable X the value 3
(since $25 \div 7 = 3$ remainder 4, so 25 DIV 7 = 3).

The MOD operation provides the **remainder**. For example,

> XREM: = 25 MOD 7;

assigns the INTEGER variable XREM the value 4.

> Y: = 100 DIV 20;   assigns Y the value 5

> YREM: = 100 MOD 20;   assigns YREM the value 0

Remember also, that integer values are limited to $\pm$MAXINT where MAXINT is the predefined constant whose value is the largest integer that can be held in the computer:

$$\text{MAXINT} = (2^{n-1} - 1) \text{ where } n = \text{word length in bits}$$

and is typically 32 767 for many micros and over 21 000 million for a 64-bit mainframe. If this range is exceeded, program execution will stop and you will be notified of "overflow error" (see also section 2.6).

Example 3.3 This program illustrates the action of the basic operations that can be performed with INTEGERs.

```
PROGRAM INTCALCS;

{ ILLUSTRATION OF SOME INTEGER CALCULATIONS }

VAR  C,D  :INTEGER;

BEGIN
  WRITELN('MAX INTEGER = ',MAXINT);
  C:=42;D:=15;
  WRITELN('C+D = ',C+D:6);
  WRITELN('CxD = ',C*D:6);
  WRITELN('C/D = ',C/D:6:3);
  WRITELN('C DIV D = ',C DIV D:6);
  WRITELN('C MOD D = ',C MOD D:6)
END.
```

On running the program, we obtain the output:

```
MAX INTEGER = 32767
C+D =         57
CxD =        630
C/D =     2.800
C DIV D =        2
C MOD D =       12
```

Example 3.4 This program may be used to convert a decimal number in the range 0 to 15 into its 4-bit binary number equivalent. It uses the DIV and MOD operations to effect the conversion. (Note that this is a very simple program; a general program for decimal-to-binary conversion and vice-versa is given in section 6.3, see example 6.3.)

```
PROGRAM DECTOBINARY;

VAR X0,X1,X2,X3,N,DECNO :INTEGER;

BEGIN
  WRITE('ENTER NUMBER ');
  READ(N);
  X0:=      N MOD 2; DECNO:=      N DIV 2;
  X1:=DECNO MOD 2; DECNO:=DECNO DIV 2;
  X2:=DECNO MOD 2; DECNO:=DECNO DIV 2;
  X3:=DECNO MOD 2; DECNO:=DECNO DIV 2;
  WRITE('DECIMAL NUMBER ',N:3,' = ',X3:1,X2:1,X1:1,X0:1);
  WRITE(' IN BINARY')
END.
```

Try running the program to see that it produces the correct results. Can you modify it to handle numbers in the range 0 to 63?

3.4 Comparison operators and introduction to decision making

The equality-inequality operators are very often used in programs for comparing two data values and, on the basis of the comparison, selecting a given course of action. There are six operators in all and they can be used with REAL, INTEGER and CHAR data. However, both members in a comparison expression must be of the same type. Mixed types cannot normally be compared.

The symbols and explanations of the meaning of the comparison operators are given on p. 43; A and B are identifiers representing data values of the same type.

It should be emphasized that comparison expressions such as $A > B$, $B <= A$, etc. always give a Boolean result, i.e. either TRUE or FALSE.

Comparison expressions must be used with care with REALs, especially for equality. REALs are stored in the computer to a limited number of significant figures (see also section 2.6) owing to finite word length. This places a restriction on their absolute accuracy and therefore two real numbers cannot always be safely compared, e.g. if A:=2·34178 and B:=2·34177 then the comparison A = B would yield a FALSE result, even though such a small difference might well be perfectly satisfactory in a particular problem solution.

Comparison expressions are extensively used as test conditions in conjunction with decision-making and control-type statements considered fully in the next chapter. The example given below employs one of these:

IF   (comparison expression—the "test" condition)

THEN   (statement(s) for one course of action)

ELSE   (statement(s) for a second course of action);

The result yielded by the comparison expression, i.e. whether its value is TRUE or FALSE, determines the subsequent course of action. IF the comparison expression is TRUE THEN first course of action ELSE (if comparison expression is FALSE) second course of action.

Example 3.5   This program simulates a resistance test on a conductor (e.g. a powerline, lead, etc.). If the conductor resistance is less than or equal to 0·25 Ω, i.e. IF R<=RLIMIT, then we are told we are within specification, ELSE we are told how far above spec. the resistance is.

```
PROGRAM RTEST;

CONST  RLIMIT=0.25;

VAR    R     :REAL;

BEGIN
 WRITE('ENTER RESISTANCE MEASURED ');
 READ(R);
  IF R <= RLIMIT THEN
   BEGIN
    WRITE('CONDUCTOR RESISTANCE IS ');
    WRITELN( 0.25-R:6:4,' WITHIN SPECIFICATION')
   END
  ELSE
   BEGIN
    WRITE('CONDUCTOR RESISTANCE IS ',R-0.25:6:4);
    WRITELN(' ABOVE SPEC. OF 0.25 OHMS')
   END
END.
```

3.5 Boolean operators and expressions

Boolean expressions are used essentially for making decisions within a program. A Boolean expression (just like a comparison expression) can take only one of two values: TRUE or FALSE.

Symbol	Meaning
=	Equality; e.g. the expression $A = B$ checks whether the value of the left-hand term A is equal to the value of the right-hand term B.
<>	Inequality; e.g. $A <> B$ checks whether A is unequal to B.
<=	Less than or equal to; e.g. $A <= B$ checks whether A is less than or equal to B.
>=	Greater than or equal to; e.g. $A >= B$ checks whether A is greater than or equal to B.
<	Less than; e.g. $A < B$ checks whether A is less than B.
>	Greater than; e.g. $A > B$ checks whether A is greater than B.

In addition to the comparison operators, four Boolean or logic operators:

NOT, AND, OR, XOR

are used to create Boolean expressions. Their meaning is explained below:

NOT the logical NOT or logical negation
e.g. NOT A is TRUE if A is FALSE,
NOT A is FALSE if A is TRUE.

AND the logical AND
e.g. A AND B is TRUE if and only if A and B are both TRUE; if A and/or B are FALSE, A AND B is assigned a FALSE value.

OR the logical OR
e.g. A OR B is TRUE if either or both A and B are TRUE; if A and B are both false, A OR B is assigned a FALSE value.

XOR the logical EXCLUSIVE OR
e.g. A XOR B is TRUE if either A or B is TRUE; if A and B are both TRUE or both FALSE, A XOR B is FALSE.

Precedence The order of precedence of these operators in evaluating Boolean expressions is as follows:

highest NOT
AND
OR, XOR
= <> <= >= < >

A simple Boolean expression consists of a series of Boolean values separated by AND, OR, XOR or preceded by NOT,

For example, suppose A, B, C, D are declared as BOOLEAN variables and assigned either TRUE or FALSE values, then *a*) A AND B AND C AND D is TRUE if and only if A, B, C, D are all assigned TRUE values.

b) A AND NOT D is TRUE if A is TRUE and D is FALSE.

c) A OR C OR D is TRUE if one or more of the variables is assigned TRUE.

d) C XOR D is TRUE if either C or D is TRUE, otherwise it is FALSE.

When it is required to combine comparison expressions or indicate precedence, ( ) type brackets must be used,

For example, if X and Y are INTEGER variables, then the Boolean expression:

(X>10) OR (Y<=100)

is assigned a TRUE value if the value of X is greater than 10 OR the value of Y is less than or equal to 100. (Note the comparison expressions must be enclosed in brackets.)

Example 3.6 The logic circuit of fig. 3.2 consisting of AND, OR and NOT gates performs the exclusive-OR function, i.e. the output F is a 1 state if either A or B is a 1 state. The Boolean expression for F is

F = A AND NOT B OR B AND NOT A

The following program allows you to check the exclusive-OR function. Try running the program with A = 0, B = 0; A = 0, B = 1; A = 1, B = 0; and A = 1, B = 1; and check the truth table also given in fig. 3.2. The program utilizes the IF · · · THEN statement (considered in detail in the next chapter). Its meaning is self-evident, e.g.

IF NA = 1 THEN A:=TRUE;

If the INTEGER variable NA is a 1 THEN the Boolean variable A is assigned a TRUE value.

```
PROGRAM LOGIC1;

VAR    A,B,F:BOOLEAN;
       NA,NB:INTEGER;

BEGIN
A:=FALSE; B:=FALSE;
 WRITE('ENTER STATES FOR A and B ');
 READ(NA,NB);
  IF NA=1 THEN A:=TRUE;
  IF NA=0 THEN A:=FALSE;
  IF NB=1 THEN B:=TRUE;
  IF NB=0 THEN B:=FALSE;
F:=A AND NOT B OR B AND NOT A;
  IF F=TRUE THEN
   WRITELN('OUTPUT, F = 1')
   ELSE
   WRITELN('OUTPUT, F = 0')
END.
```

Fig. 3.2 Exclusive-OR circuit and truth table

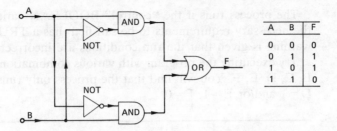

A	B	F
0	0	0
0	1	1
1	0	1
1	1	0

Example 3.7 This program simulates a simple thermostat control where a heater and/or boost and/or fan is turned on or off. It employs comparison operators, the logical AND and IF...THEN statements.

```
PROGRAM CONTROL1;

{ SIMULATION OF THERMOSTATIC CONTROL }

CONST   HITEMP=25; LOWTEMP=15;

VAR     TEMP:REAL;

BEGIN
 WRITE('ENTER TEMPERATURE '); READ(TEMP);
  IF TEMP<=LOWTEMP THEN
    WRITELN('SWITCH ON HEATER + BOOST');
  IF (TEMP>LOWTEMP) AND (TEMP<HITEMP) THEN
    WRITELN('HEATER ON, SWITCH OFF BOOST');
  IF TEMP>=HITEMP THEN
    BEGIN
      WRITELN('SWITCH OFF HEATER');
      WRITELN('SWITCH ON FAN')
    END
END.
```

Example 3.8 This program can be regarded as a simple example of a control program for an industrial process governed by the following conditions:

the process runs if

A   the start button is "on"

AND B   the flow of material is sufficient

AND C   the temperature is high enough

AND D OR E   either one or other of two other conditions is O.K.

AND NOT F   the emergency stop button is not pressed

In the program execution these conditions are first read, i.e. the INTEGER variables A, B, C, D, E, F simulate the process input information to the computer. We then define corresponding BOOLEAN variables SA, SB, SC, SD, SE, SF which take either TRUE or FALSE values on the basis of this input information, i.e. if A = 1 then SA is assigned a TRUE value, if A = 0 then SA is FALSE, and so on.

The process runs if the "control" BOOLEAN variable Q (containing all the necessary requirements to be satisfied) has a TRUE value. Otherwise a warning is given that the run conditions are incorrect.

Try executing the program with various combinations of 1–0 values for A, B, C, D, E, F. You will find that the process only runs when A = B = C = 1, D = 1 and/or E = 1, F = 0.

```
PROGRAM CONTROL2;
{ SIMULATION OF START-UP CONTROL
  FOR AN INDUSTRIAL PROCESS }

VAR A,B,C,D,E,F :INTEGER;
    SA,SB,SC,SD,SE,SF,Q :BOOLEAN;

BEGIN
  WRITELN('ENTER INPUT CONDITIONS FOR');
  WRITE('A B C D E F...1 OR 0:  ');
  READ(A,B,C,D,E,D,F);
   SA:= A=1; SB:= B=1; SC:= C=1;
   SD:= D=1; SE:= E=1; SF:= F=1;
   Q:=SA AND SB AND SC AND (SD OR SE) AND NOT SF;
     IF Q=TRUE THEN
        WRITELN('RUN PROCESS')
        ELSE
        WRITELN('START CONDITS. NOT SATISFIED')
END.
```

3.6 Standard arithmetic functions

Several standard functions are provided in PASCAL for executing prescribed tasks on REAL, INTEGER and CHAR data.

We describe here the form and meaning of those used with REAL and INTEGER data to return the value of mainly arithmetic functions. In the next section we consider standard functions involving characters.

Table 3.1 summarizes the standard arithmetic functions normally available in most implementations of PASCAL. Note the type of the argument, the X in brackets, and the type of the result obtained.

1 ABS(X)

returns the absolute value of X; if X is REAL then ABS(X) is REAL; if X is INTEGER then ABS(X) is INTEGER. Examples:

A:=ABS(−89·3);   assigns A the value +89·3
B:=ABS(−467);   assigns B the value +467
C:=ABS(3∗3+4∗4);   assigns C the value 25
D:=ABS(3∗3−4∗4);   assigns D the value 7

Table 3.1 Standard arithmetic functions

Function identifier	Meaning	Type of argument	Type of result
ABS	ABS(X) finds absolute value of X	REAL INTEGER	REAL INTEGER
TRUNC	TRUNC(X) gives whole number part of X	REAL	INTEGER
ROUND	ROUND(X) rounds X to nearest whole number	REAL	INTEGER
ODD	ODD(X) gives TRUE if X is odd else FALSE	INTEGER	BOOLEAN
SQRT	SQRT(X) gives $\sqrt{X}$	REAL or INTEGER	REAL
SQR	SQR(X) gives X^2	REAL INTEGER	REAL INTEGER
SIN	SIN(X) gives SIN X X assumed in radians	REAL or INTEGER	REAL
COS	COS(X) gives COS X X in radians	REAL or INTEGER	REAL
ARCTAN	ARCTAN(X) gives $\tan^{-1} X$	REAL or INTEGER	REAL
EXP	EXP(X) gives e^x	REAL or INTEGER	REAL
LN	LN(X) gives ln X (natural log), X>0	REAL or INTEGER	REAL

2 TRUNC(X)

returns the whole number part of X, discarding any fractional part.

TRUNC(7·829)   returns the value 7

TRUNC(−12·54)   returns the value −12

A:=TRUNC(−7·8);   assigns A the value −7

3 ROUND(X)

rounds X to the nearest whole number.

ROUND(7·923)   returns the value 8

ROUND(−2·54)   returns the value −3

ROUND(−2·49)   returns the value −2

When X is precisely 0·5 greater than an integer, ROUND(X) rounds X up when X is positive, down if X is negative, e.g.

ROUND(7·5) returns 8, ROUND(−7·5) returns −8

4 ODD(X)

tests whether or not the INTEGER value X is odd; ODD(X) is assigned TRUE if X is odd, FALSE if X is even.

P:=ODD(5)   assigns the BOOLEAN variable a TRUE value

Q:=ODD(4)   assigns Q a FALSE value

5 SQR(X)

returns the square of X.

A:=SQR(2.61);   assigns the REAL variable A the value $2 \cdot 61^2 = 6 \cdot 8121$

N:=SQR(7);   assigns the INTEGER variable N the value $7^2 = 49$

6 SQRT(X)

returns the square root of X, provided $X \geqslant 0$.

A:=SQRT(81);   assigns to the REAL variable A the value $\sqrt{81} = 9 \cdot 0$

B:=SQRT (5·0176);   assigns B the value $\sqrt{5 \cdot 0176} = 2 \cdot 24$

7 SIN(X), COS(X)

return the values of sin X and cos X respectively.

Note that X is in radians and remember

$$X \text{ radians} = X \text{ degrees} \times \frac{\pi}{180°}$$

so, for example, to find sin 42°:

WRITE(SIN(42*PI/180));

i.e. X degrees must first be converted to radians.

Note also that PASCAL does not provide a standard function for tan X. However we can write our own function to do this (see Chapter 5, section 5.6) or use the identity tan X = sin X/cos X directly in our program, e.g. to find tan 37°:

X:=37*PI/180;

WRITE('TAN 37 = ', SIN(X)/COS(X));

8 ARCTAN(X)

returns the value whose tangent is X, the result being specified in radians between $-\pi/2$ and $+\pi/2$.

A:=ARCTAN(1);

assigns A the value 0·78539816 radians, i.e. $\pi/4$ or 45°.

B:=ARCTAN(-0·72);

assigns B the value −0·6240225 radians (−35·75°).

9 EXP(X)

returns the value of the exponential function e^x, where $e = 2·718\,281\,828$.

EXP(1·29)   returns the value $e^{1·29} = 3·632\,786\,56$

EXP(−2·4)   returns $e^{−2·4} = 9.071\,795\,3E−02$

10 LN(X)

returns the value of the natural logarithm (log to base e) of X. Note that X must be greater than 0.

Example 3.9 PASCAL provides no direct means of calculating powers. This program can be used to calculate x^y. x must be greater than zero; y can be any value (within the computer range), positive, negative, fractional. The program is based on the following result:

if $P = X^Y$

then $\ln P = Y \ln X$ (on taking logs of both sides)

so $P = e^{Y\ln X}$

(as, in general, if $\log_a N = k$ then $N = a^k$).

In Chapter 5 we show how to write a user-defined function to calculate powers.

```
PROGRAM POWERCALC;

{ TO CALCULATE X TO POWER Y }

VAR X,Y,P  :REAL;

BEGIN
  WRITE('ENTER X,Y VALUES ');
  READ(X,Y);
  P:=EXP(Y*LN(X));
  WRITE('X TO POWER Y = ',P)
END.
```

Example 3.10 This program determines the impedance, phase angle and rms current for the series L–R circuit of fig. 3.3. It uses the standard functions SQRT and ARCTAN to help solve the problem.

Fig. 3.3 A.C. analysis of series L–R circuit

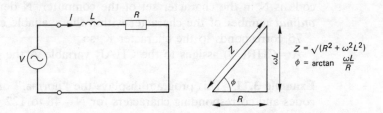

$$Z = \sqrt{(R^2 + \omega^2 L^2)}$$
$$\phi = \arctan \frac{\omega L}{R}$$

```
PROGRAM ACCIRCUITS1;

{ AC ANALYSIS OF SERIES L-R CIRCUIT }

CONST PI=3.14159; {USUALLY PREDEFINED}

VAR  L,R,FREQ,W,V,I,PHI,Z :REAL;

BEGIN
 WRITE('ENTER R AND L VALUES  ');READ(R,L);
 WRITE('ENTER RMS SOURCE VOLTAGE ');READ(V);
 WRITE('SOURCE FREQUENCY  ');READ(FREQ);
 W:=2*PI*FREQ;
 Z:=SQRT(R*R+W*L*W*L);
 I:=V/Z;
 PHI:=  ARCTAN(W*L/R)*180/PI;
 WRITELN('RESULTS FOR L-R CIRCUIT :');
 WRITELN('===========================');
 WRITELN('IMPEDANDCE   Z = ',Z:8:2);WRITELN;
 WRITELN('RMS CURRENT I = ',I:8:3);WRITELN;
 WRITELN('PHASE ANGLE   = ',PHI:8:1)
END.
```

Here is an example run, where $R = 100\,\Omega$, $L = 0.3\,\text{H}$, $V = 240\,\text{V}$ and $f = 50\,\text{Hz}$ have been entered:

```
ENTER R AND L VALUES  100 0.3
ENTER RMS SOURCE VOLTAGE 240
SOURCE FREQUENCY  50
RESULTS FOR L-R CIRCUIT :
===========================
IMPEDANCE   Z =   137.41

RMS CURRENT I =    1.747

PHASE ANGLE   =    43.3
```

3.7 Standard functions involving characters

There are four standard functions in PASCAL associated with CHARacter data.

1 CHR(N): the character function
CHR(N) where N is a positive INTEGER value gives the character whose code is N in the character set of the computer. N denotes the position or *ordinal* number of the character within the available character set, e.g.

75 corresponds to the character K, so

A:=CHR(75) assigns to the CHAR variable A the character K

Example 3.11 This program displays the "normal" or standard ASCII set codes and corresponding characters for $N = 48$ to 122 (this range is universal for all computers). We use the FOR · · · TO · · · DO loop to effect the display; the FOR loop will be explained in the next chapter.

```
PROGRAM CODECHARSET;

( TO DISPLAY CODE CHARACTER SET
          OVER RANGE 48 TO 122}
VAR C:CHAR;
    N:INTEGER;

BEGIN
 WRITELN(' CODE     CHARACTER');
  FOR N:=48 TO 122 DO
    WRITELN(N:4,CHR(N):9)
END.
```

On running the program, you obtain a tabulation of code (ordinal number) and its corresponding character as shown below:

CODE	CHARACTER		CODE	CHARACTER
48	0		86	V
49	1		87	W
50	2		88	X
51	3		89	Y
52	4		90	Z
53	5		91	[
54	6		92	\
55	7		93	]
56	8		94	^
57	9		95	_
58	:		96	`
59	;		97	a
60	<		98	b
61	=		99	c
62	>		100	d
63	?		101	e
64	@		102	f
65	A		103	g
66	B		104	h
67	C		105	i
68	D		106	j
69	E		107	k
70	F		108	l
71	G		109	m
72	H		110	n
73	I		111	o
74	J		112	p
75	K		113	q
76	L		114	r
77	M		115	s
78	N		116	t
79	O		117	u
80	P		118	v
81	Q		119	w
82	R		120	x
83	S		121	y
84	T		122	z
85	U			

2 ORD(C): the ordinal function

ORD(C) where C must be a CHAR value gives the ordinal number of the value within the character set, e.g. if C is assigned the value ? then ORD(C) would return 63, the ordinal number of code for ?, i.e.

C:='?';

WRITE(ORD(C));

would cause 63 to be displayed.

CHR and ORD are essentially *inverse* functions: the CHR function gives the character for the code (ordinal number), the ORD function gives the code for the character.

3 PRED(C): the predecessor function

PRED(C) gives the previous character to the CHAR value C in the character set, e.g.

if C is assigned 'M', then PRED(C) returns L

4 SUCC(C): the successor function

SUCC(C) gives the character following the CHAR value C in the character set, e.g.

if C is assigned X, then SUCC(C) returns Y

PRED and SUCC are also inverse functions.

Exercises 3

3.1 Using the WRITE statement, construct PASCAL programs to evaluate the following

a) $\sqrt{(5 \cdot 6 \times 2 \cdot 3)}$ b) $3 \cdot 8(5 - 0 \cdot 65 \times 2 \cdot 1)$

c) $\dfrac{5 \cdot 6 \times 2 \cdot 3}{10 \cdot 2 - (9 \cdot 4 \times 0 \cdot 7)}$ d) $\sqrt{\left\{ \dfrac{69 \cdot 8}{(12 \cdot 67 - 2 \cdot 98)^2} \right\}}$

3.2 Explain the meaning of the integer arithmetic operators DIV and MOD and evaluate the following:

16 DIV 3   8 MOD 5   (14 MOD 5) DIV 2

3.3 Given that 1 inch = 25·4 mm write a program that will read feet and inches from the keyboard, convert the feet and inches to metres, and output the result to the screen.

3.4 Explain the meaning of the following standard functions: ABS, TRUNC, ROUND, ARCTAN, CHR, ORD, SUCC; and evaluate the following:

ABS(−7·0)   TRUNC(14·6)   TRUNC(−14·6)

ROUND(−14·6)   ARCTAN(−1·0)   CHR(97)

ORD('?')   SUCC('D')

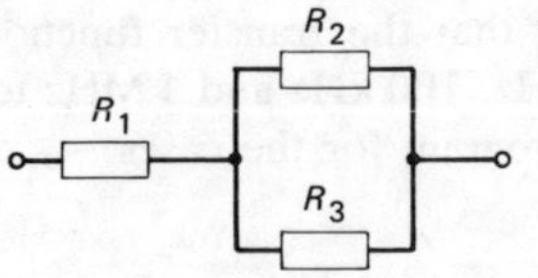

Fig. 3.4

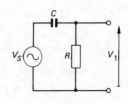

3.5 Write a program that will determine the total resistance of the resistive circuit of fig. 3.4. The resistor values are to be entered from the keyboard and the total resistance is to be output to the V.D.U.

3.6 Modify the program of **3.5** so that both resistance values and their tolerances (e.g ±10%, ±20%) are entered in and the maximum and minimum total circuit resistances that might be expected are output to the V.D.U.

3.7 Fig. 3.5 shows a simple bias circuit for a bipolar transistor. The values of the base resistor R_B and collector load resistor R_C in terms of collector power supply V_{CC} and d.c. (quiescent) bias point values are given by

$$R_C = \frac{V_{CC} - V_{CE}}{I_C} \qquad R_B = \frac{V_{CC} - V_{BE}}{I_B}$$

The collector and base current are related by

$$I_C = h_{FE} I_B \qquad h_{FE} = \text{d.c. static current gain}$$

and the base-emitter voltage V_{BE} can be taken as approximately constant and equal to 0·65 V.

  Write a general program to evaluate R_C and R_B for quiescent point values (V_{CE}, I_C) and V_{CC} and h_{FE} values input from the keyboard. Run your program to determine R_C and R_B for

a) $V_{CC} = 20$ V, $h_{FE} = 100$, $V_{CE} = 8$ V, $I_C = 1$ mA

b) $V_{CC} = 12$ V, $h_{FE} = 50$, $V_{CE} = 6$ V, $I_C = 4$ mA

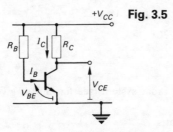

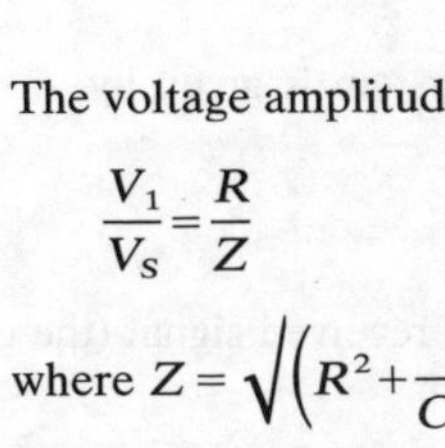

Fig. 3.5

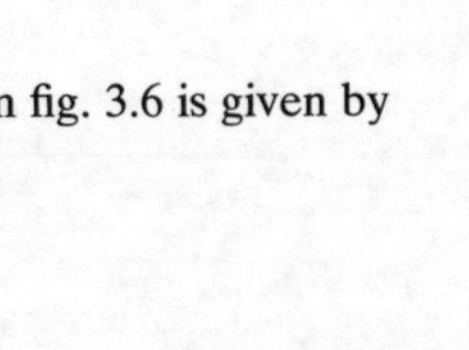

Fig. 3.6

3.8 The voltage amplitude transfer function for the circuit shown in fig. 3.6 is given by

$$\frac{V_1}{V_S} = \frac{R}{Z}$$

where $Z = \sqrt{\left(R^2 + \dfrac{1}{C^2\omega^2}\right)}$ is the impedance of the C-R network. Write a program to calculate and display the value of the transfer function for values of C, R and f (f = frequency, $\omega = 2\pi f$) input from the keyboard.

3.9   Modify the program of **3.8** so that the transfer function values are displayed for $f = 10$ Hz, 100 Hz, 1 kHz, 10 kHz, 100 kHz and 1 MHz for values of C and R input from the keyboard. Run the program for the cases
   a) $C = 0{\cdot}1\ \mu\mathrm{F}$, $R = 1\ \mathrm{k\Omega}$
   b) $C = 10\ \mu\mathrm{F}$, $R = 1\ \mathrm{k\Omega}$

Fig. 3.7 Line-of-sight communications link

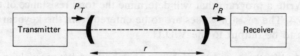

3.10   In a line-of-sight communications link (see fig. 3.7) the total power received is given by

$$P_R = \frac{\lambda^2 G_T G_R}{16\pi^2 r^2} P_T \text{ watts}$$

where λ = wavelength of transmission; $\lambda = (3 \times 10^8)/f$, f = frequency
   G_T = gain of transmitter antenna
   G_R = gain of receiver antenna
   P_T = total power transmitted
   r = range, distance between transmitter and receiver
Write a program to evaluate P_R and the transmission loss $10 \log (P_T/P_R)$ decibels with the systems data (frequency, antenna gains, range, transmitter power) being input from the keyboard.

   Run your program for the case of an $f = 3$ GHz link, employing transmitter and receiver antennas both of gain 100, transmitting a power of 4 W over a range of 50 000 m.

Fig. 3.8 Radar system

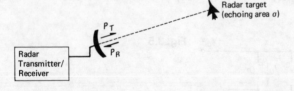

3.11   The power received by a radar system (see fig. 3.8) is given by

$$P_R = \frac{\sigma G_T G_R \lambda^2}{64\pi^3 r^4} P_T \text{ watts}$$

The maximum range of the radar system is given by

$$r_{max} = \sqrt[4]{\left[\frac{\sigma G_T G_R \lambda^2 P_T}{64\pi^3 P_{R\ min}} \right]} \text{ metres}$$

where $P_{R\ min}$ is the smallest possible received signal (the echo) which the system can recognise and process without error.
   $P_{R\ min}$ is given by

$$P_{R\ min} = kB[T_s + (F-1)T_0](S_0/N_0)_{min}$$

where $k = 1{\cdot}38 \times 10^{-23}$ J/K, $T_0 = 290$ K

B = receiver bandwidth in hertz

F = receiver noise factor

T_s = receiver antenna noise temperature, K

σ = echo area of target, m^2

G_T, G_R = gain of transmitter, receiver antenna (normally common)

λ = transmitter wavelength, m

P_T = total transmitter power, W

r = target range, m

$(S_0/N_0)_{min}$ = minimum signal-to-noise ratio at receiver for satisfactory detection.

Write the following programs:

a) To evaluate received power P_R inputting data for σ, G_R, G_T, λ, P_T and r from the keyboard.

b) To find the maximum range of a system for the case

Wavelength = 3 cm

Peak transmitter power = 80 kW

Antenna and receiver gain = 500

Antenna noise temperature = 290 K

Receiver noise factor = 10

Receiver bandwidth = 5 MHz

Target echoing area = 60 m^2

$(S_0/N_0)_{min} = 4$

4 Statements for Choice, Selection and Repetition

4.1 Introduction and summary

PASCAL provides directly for three important requirements which are invariably needed in the solution of problems:

CONDITIONAL CHOICE: choice of one or an alternative course of action.
SELECTION: selection of one of many alternative courses of action.
REPETITION: repeated execution of a section of program statements

In this chapter we consider the PASCAL structured statements used to exercise these forms of control. We explain and give practical program examples for

CONDITIONAL CHOICE: the IF · · · THEN and
$\qquad\qquad\qquad\qquad$ IF · · · THEN · · · ELSE statements
SELECTION: the CASE statement

REPETITION:
1. WHILE some condition is satisfied, the WHILE · · · DO statement
2. UNTIL some condition is satisfied, the REPEAT · · · UNTIL statement
3. FOR a given number of times, the FOR · · · TO · · · DO statement

and finally we consider the GOTO jump statement.

4.2 The IF statements

In PASCAL the choice of one of two different courses of action can be made by the **IF · · · THEN · · · ELSE** structured statement. The syntax diagram for this statement is shown in fig. 4.1.
The form of the statement is:

    IF (test expression) THEN (Action 1) ELSE (action 2);

Note
1. The expression following IF acts as a test condition (a comparison or Boolean type expression) and yields either a TRUE or FALSE value.

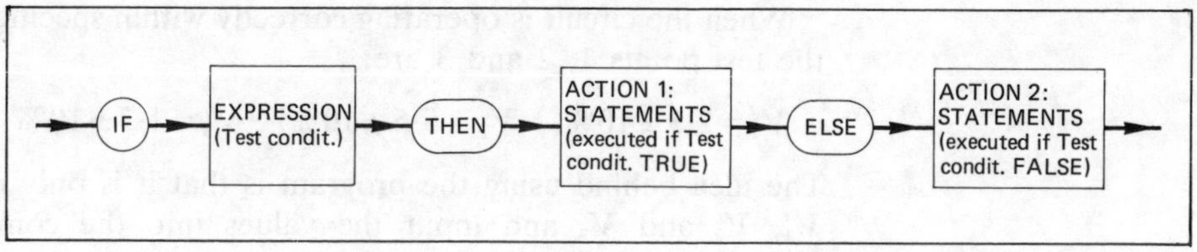

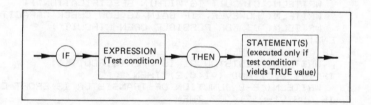

Fig. 4.1 Syntax diagram for **IF... THEN ...ELSE**

2. IF the test expression yields a TRUE value THEN the statement(s) of action 1 are executed.
3. ELSE (if the test condition yields a FALSE value) the statement(s) of action 2 are executed.
4. If either action consists of a number of statements, use BEGIN and END to bracket them into a compound statement.

When it is required to execute an action only if a certain condition is satisfied (i.e. no alternative action required), then the IF··· THEN statement is used. The syntax diagram for this statement is shown in fig. 4.2. If the test expression after IF yields a TRUE value, the statement(s) following THEN are executed; if the value is FALSE, the statement(s) are skipped over and execution proceeds to the next program section.

Fig. 4.2 Syntax diagram IF... **THEN**

We have already used the IF statements as a means of introducing the use of comparison and Boolean operators in creating "test" expressions; see examples 3.5, 3.6 and 3.7. Here is one more.

Example 4.1 This program gives an example of a computer-aided fault diagnosis routine for checking the performance of the common emitter amplifier circuit shown in fig. 4.3.

Fig. 4.3 Single-stage transistor amplifier: fault diagnosis

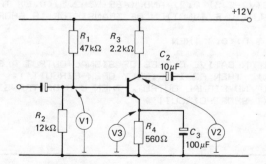

When the circuit is operating correctly within specification, the voltages at the test points 1, 2 and 3 are:

$$V_1 = 2 \cdot 3 \pm 10\%, \quad V_2 = 5 \cdot 5 \pm 10\%, \quad V_3 = 1 \cdot 7 \pm 10\%$$

The idea behind using the program is that it is only necessary to measure V_1, V_2 and V_3 and input the values into the computer. The program displays whether or not the circuit is operating within specification, and if not what the fault is.

```
PROGRAM AMPTEST1;

VAR    V1,V2,V3 :REAL;

BEGIN
 WRITELN('ENTER VOLTAGES V1,V2,V3 MEASURED');
 READ(V1,V2,V3);
 WRITELN;
 IF (ABS(V1-2.3)<0.2) AND (ABS(V2-5.5)<0.5)
      AND (ABS(V3-1.7)<0.2) THEN
  BEGIN
    WRITELN('CIRCUIT IS WITHIN SPECIFICATION');
    WRITELN('HOWEVER, IF GAIN IS LOW CHECK CAPACITOR');
    WRITELN('C3 FOR POSSIBLE OPEN-CIRCUIT')
  END;
 IF  V1=0.0  THEN
    WRITELN('RESISTOR R1 IS OPEN-CIRCUIT');
 IF (V1>0.0) AND (V1<0.2) THEN
    WRITELN('E-B JUNCTION OF TRANSISTOR IS SHORT-CIRCUIT');
 IF ABS(V1-0.7)<0.2 THEN
    BEGIN
     IF V2<0.5 THEN
       WRITELN('RESISTOR  R3 IS OPEN-CIRCUIT OR C3 SHORT-CIRCUIT');
     IF ABS(V2-12)<2.0 THEN
        WRITELN('C-B JUNCTION OF TRANSISTOR IS OPEN-CIRCUIT');
    END;
 IF ABS(V1-2.3)<0.2 THEN
    BEGIN
     IF V3< 0.05 THEN
       WRITELN('E-B JUNCTION OF TRANSISTOR IS OPEN-CIRCUIT');
     IF( ABS(V2-12)<2.0) AND (ABS(V3-2)<0.2) THEN
       WRITELN('RESISTOR R4 IS OPEN-CIRCUIT');
     IF (ABS(V2-2.5)<0.2) AND (ABS(V3-2.5)<0.2) THEN
       WRITELN('C-E JUNCTION OF TRANSISTOR IS SHORT-CIRCUIT')
    END;
 IF ABS(V1-3.0)<0.4 THEN
   BEGIN
    WRITE('IF NEGATIVE CYCLES OF SIGNAL OUTPUT ALSO CLIPPED');
    WRITELN(', THEN R2 IS LIKELY OPEN-CIRCUIT');
    WRITE('IF NO SIGNAL OUTPUT, THEN C-B OF TRANSISTOR IS');
    WRITELN(' SHORT-CIRCUIT')
   END
END.
```

4.3 The CASE statement

The IF statements allow only the choice between two courses of action. Frequently, however, we require the facility to select one of several alternatives. For this, PASCAL provides the **CASE** structured statement, whose syntax diagram is given in fig. 4.4.

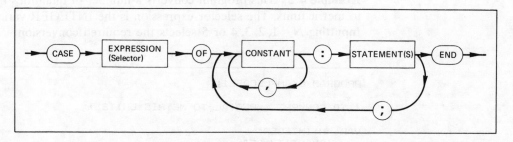

Fig. 4.4 Syntax diagram for **CASE statement**

The expression after CASE acts as the selector. Its value determines which of the alternative courses is executed. Each alternative is identified by a CONSTANT, know as the **case label**. The selector expression and the case label constant must be of the same type: INTEGER, CHAR, BOOLEAN or user-defined but not REAL. For example,

CASE N OF 1: statement(s) 1;
      2: statement(s) 2;
      3: statement(s) 3;
Selector   └── Case label constant

The selector expression in this example is simply the INTEGER variable N. If N = 1, THEN statement(s) 1 following 1: are executed. If N = 2, then statement(s) 2 following 2: are executed ... and so on.

CASE LETTER OF 'A': statement(s) A;
       'B': statement(s) B;
       'C': statement(s) C;
       'D': statement(s) D;

In this case the selector expression is the CHAR variable LETTER. When LETTER = 'A', then statement(s) A are executed. If LETTER = 'B', statement(s) B are executed...and so on.

Two or more case labels, separated by commas, can also be used to direct execution to a given course of action, e.g.

CASE RVALUE OF
80, 81, 82, 83, 84, 85: statement(s) 1;
100, 120, 140   : statement(s) 2;
150, 200, 470, 560 : statement(s) 3;

Thus if RVALUE has the value 80, 81, 82, 84 or 85, statement(s) 1 are executed; if RVALUE is 100, 120 or 140, statement(s) 2 are executed; if RVALUE is 150, 200, 470 or 560, statement(s) 3 are executed.

In standard PASCAL, the selector expression must take one of the values specified by the case label constants. If it evaluates to a value not specified, then a run-time error is displayed and program execution stops. To avoid this, always ensure that the selector expression can only take values which are specified as case labels.

Example 4.2 This program converts a number of quantities from Imperial to metric units. The selector expression is the INTEGER variable value N. Inputting $N = 1, 2, 3, 4$ or 5 selects the required conversion.

```
PROGRAM EXAMPLECASE1;

{ TO CONVERT IMPERIAL TO METRIC UNITS }

VAR   N:INTEGER;
      QUANTITY:REAL;

BEGIN
 WRITELN('TO CONVERT :');
 WRITELN('POUNDS TO KILOGRAMMES......ENTER 1');
 WRITELN('FEET TO METRES.............ENTER 2');
 WRITELN('M.P.H. TO m/s..............ENTER 3');
 WRITELN('GALLONS TO LITRES..........ENTER 4');
 WRITELN('HORSE POWER TO WATTS.......ENTER 5');
 READ(N);
 WRITE('NOW ENTER QUANTITY TO BE CONVERTED ');
 READ(QUANTITY);
  CASE N OF
      1:WRITELN('NO. OF KILOS = ',QUANTITY*0.45359:10:3);
      2:WRITELN('NO. OF METRES = ',QUANTITY*0.3048:10:3);
      3:WRITELN(QUANTITY*0.44704:10:3,' m/s');
      4:WRITELN('NO. OF LITRES = ',QUANTITY*4.546:10:3);
      5:WRITELN(QUANTITY:8:3,' H.P. = ',QUANTITY*745.7:10:3,' W')
  END
END.
```

4.4 Repetition 1: the WHILE · · · DO statement

PASCAL provides two forms of loop-type statements which permit the repeated execution of a section of program statements to be made conditional on the value of a Boolean type expression. These are the WHILE · · · DO and REPEAT · · · UNTIL statements:

WHILE (Boolean Expression) DO (Action);

REPEAT (Action) UNTIL (Boolean Expression);

The syntax diagram for the **WHILE · · · DO** statement is shown in fig. 4.5, whilst a flowchart representation of its action is given in fig. 4.6. The test expression following WHILE controls whether or not the statement(s) following DO are executed. This expression is evaluated at the beginning of each cycle, rather than at the end, which is the case for the REPEAT · · ·

Fig. 4.5 Syntax diagram for **WHILE. . . DO statement**

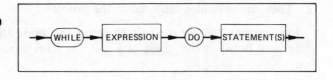

Fig. 4.6 Flowchart representation of WHILE. . . DO loop

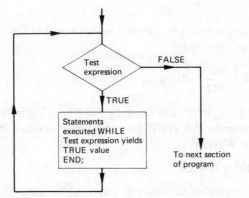

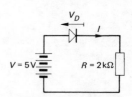

Fig. 4.7 Diode circuit

UNTIL loop. If the expression yields a TRUE value before entering the WHILE loop, the statement or, more usually, the compound statement following DO is executed. The looping process continues so long as this expression yields TRUE. As soon as the expression evaluates to FALSE, the statement(s) following DO are skipped and execution jumps to the next part of the program. Thus you must always remember to include at least one statement in this group which alters the value of test expression to provide means of exit, otherwise the loop executes forever.

Example 4.3 Determine the voltage across the diode and the current flowing in the circuit of fig. 4.7 given the current-voltage relation for the diode is

$$I_D = I_0(e^{\lambda V_D} - 1)$$

where $I_0 = 10^{-8}$ A, $\lambda = 38 \cdot 6$ V^{-1}.

Since the circuit contains a non-linear element (the diode), we cannot solve the problem using straightforward algebraic analysis. Instead we are going to make a sensible guess for the current I and then use the WHILE $\cdots$ DO loop to refine this guess until we obtain I accurate to, say, $0 \cdot 2$ mA.

Our approach is as follows:

1 First we define the constants in our program, i.e.

   Supply voltage V = 5

   Diode data: I0 = 1E-05 mA   (as $I_0 = 10^{-8}$ A)

   L = 38·6   (for $\lambda = 38 \cdot 6$)

   R = 2   (as $R = 2$ kΩ)

2 Then the variables required in the problem:

I for current, VD for the voltage across the diode and DI to represent the difference between our guess of I and the value evaluated using the diode law

$$I_D = I_0(e^{\lambda V_D} - 1)$$

3 The main program is initiated with a sensible first guess of V_D, e.g. if V_D is first set at zero, then

$$I = \frac{V}{R} = \frac{5 \text{ volts}}{2 \text{ k}\Omega} = 2 \cdot 5 \text{ mA}$$

and $\quad DI = I - I_D = 2 \cdot 5 \text{ mA} \quad$ (as $I_D = 0$ when $V_D = 0$).

4 We then enter the WHILE · · · DO loop and ask for repeated calculation to be made WHILE

$$ABS(DI >= 0 \cdot 2)$$

When this degree of accuracy is achieved we exit the WHILE · · · DO loop and display the results.

Note: exercise great care in selecting your test condition accuracy and the increments (or decrements) of your guess. The exponential function changes very rapidly with its argument (in this case L * VD) and so if you choose a very tight accuracy you may need very fine decrements in I, since I determines VD through the relation VD = V − R * I. If not, you may never exit the WHILE loop and/or an over-flow error will be registered. You simply will never obtain a FALSE value for your test condition. Can you improve the present program to avoid this possibility?

```
PROGRAM DIODECALC;

CONST    IO=1E-05 ;{milliamps.}
         L=38.6;
         V=5;
         R=2 ;{kilohms}

VAR      I,DI,VD     :REAL;

BEGIN
  I:=2.5;DI:=2.5;VD:=0.0;
   WHILE ABS(DI)>=0.2 DO
    BEGIN
      I:=I-0.01 ;
      VD:=V-R*I;
      DI:=I-IO*(EXP(L*VD)-1.0)
    END;
  WRITELN('VD= ',VD:8:3,' V');
  WRITELN('I=  ',I:8:3,' mA');
  WRITELN('DI= ',DI:8:3,' mA')
END.
```

On running the program, the solution obtained is

```
VD=    0.320 V
I=     2.340 mA
DI=    0.027 mA
```

4.5   Repetition 2: the REPEAT···UNTIL statement

The syntax diagram and a flowchart representation for the **REPEAT···
UNTIL** statement are shown in figs. 4.8 and 4.9, respectively.

Fig. 4.8 Syntax diagram for **REPEAT... UNTIL statement**

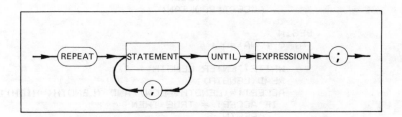

Fig. 4.9 Flowchart representation of REPEAT... UNTIL loop

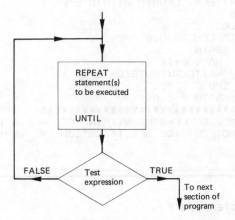

The REPEAT construct causes the statement(s) grouped between REPEAT and UNTIL to be repeatedly executed until the expression immediately following UNTIL yields a TRUE value. These statements are executed at least once and must contain at least one statement that has an effect on the test expression to allow exit. Otherwise the repetition will continue forever.

Example 4.4   This example simulates an acceptance test and count procedure. Values of LENGTH are entered in from the keyboard, "tested" to be within ±1% of 100 and the numbers within specification and outside counted. The process is terminated by entering a value of 0 or any negative value. (The program is on p. 64.)

Example 4.5   This program uses the REPEAT statement to count the number of characters in a sentence entered from the keyboard. Note that the count will include any spaces and the final full-stop. (Program on p. 64.)

```
PROGRAM TESTCOUNT;

CONST  LOW=99; HIGH=101;

VAR    LENGTH:REAL;
       A,NA  :INTEGER;
       ACCEPT:BOOLEAN;

BEGIN
 A:=0 ; NA:=0 ;
   REPEAT
     WRITE('ENTER LENGTH : ');
     READ(LENGTH);
     ACCEPT:=(LENGTH>=LOW) AND (LENGTH<=HIGH);
      IF ACCEPT = TRUE THEN
          BEGIN
           A:=A+1;
           WRITELN('LENGTH WITHIN SPEC')
          END
          ELSE
            IF LENGTH>0.0 THEN
             BEGIN
              NA:=NA+1;
              WRITELN('OUTSIDE SPEC')
             END
     UNTIL LENGTH <=0.0;
 WRITELN('***************************');
 WRITELN('NO. OF LENGTHS WITHIN SPEC = ',A);
 WRITELN('NO. OUTSIDE SPECIFICATION  = ',NA)
END.
```

```
PROGRAM LETTERCOUNT;

VAR LETTER :CHAR;
    N:     INTEGER;

BEGIN
 N:=0;
 WRITELN('TYPE IN YOUR SENTENCE');
  REPEAT
   N:=N+1;
   READ(LETTER)
  UNTIL LETTER='.';
 WRITELN('NO OF CHARACTERS = ',N)
END.
```

4.6 Repetition 3: the FOR · · · TO · · · DO statement

Both the WHILE · · · DO and REPEAT · · · UNTIL constructs are conditional types of repetition which depend on a test condition, and therefore the number of times that the loop is executed is not known. When the number of repetitions is known or can be calculated in advance, PASCAL provides the **FOR** statements:

FOR(identifier:=expression-1) TO (expression-2) DO (action)

and

FOR(identifier:=expression-1) DOWNTO (expression-2) DO (action)

The first is the incremental form, the second the decremental form.

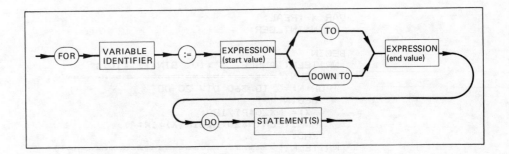

Fig. 4.10 Syntax diagram for **FOR statement**

The syntax diagram for the FOR statement is shown in fig. 4.10. The identifier after FOR is known as the **control variable** and must be of ordinal data type (i.e. the data values are "ordered", occur in sequence, are "countable" and may be compared with each other). Thus INTEGER, CHAR and user-defined "enumerated" type (see section 2.13) may be used as control variables but REALs are not allowed. The start and end values (expression 1 and expression 2) must be of the same type as the control variable. When the start value is less than the end value, TO applies; when greater, DOWNTO applies. For example

1 FOR N:=1 TO 10 DO
    WRITE(N:3);
would produce the display:

   1 2 3 4 5 6 7 8 9 10

2 FOR N:=10 DOWNTO 1 DO
    WRITE(N:3);
    would produce the display:

10 9 8 7 6 5 4 3 2 1

3 FOR LETTER:='A' TO 'M' DO
    WRITE(LETTER:2);
    would produce the display:

    A B C D E F G H I J K L M

4 FOR LETTER:='Z' TO 'N' DO
    WRITE(LETTER:2);
    would produce the display:

Z Y X W V U T S R Q P O N

Example 4.6 This example uses the FOR statement to produce a table of
sin x values for $x = 0°$ to $360°$ in steps of $20°$.

```
PROGRAM SINETABLE;

CONST PI=3.14159;

VAR X :REAL;
    N: INTEGER;

BEGIN
 WRITELN('ANGLE, X':10,'SIN(X)':10);
 WRITELN('=================================');
  FOR N:=0 TO 360 DIV 20 DO
    BEGIN
      X:= N*20*PI/180;
      WRITELN(N*20:6,SIN(X):14:4)
    END;
 WRITELN('=================================')
END.
```

On running the program the table displayed is:

```
ANGLE, X     SIN(X)
=================================
    0        0.0000
   20        0.3420
   40        0.6428
   60        0.8660
   80        0.9848
  100        0.9848
  120        0.8660
  140        0.6428
  160        0.3420
  180        0.0000
  200       -0.3420
  220       -0.6428
  240       -0.8660
  260       -0.9848
  280       -0.9848
  300       -0.8660
  320       -0.6428
  340       -0.3420
  360       -0.0000
=================================
```

Example 4.7 This example uses the FOR statement to perform repeated
calculations to find the d.c. transient response of the $R–C$ circuit shown in
fig. 4.11.

The analytical expression for the voltage v_C across the capacitor C as a
function of time t is

$$v_C = E + (V_0 - E)e^{-t/CR}$$

Fig. 4.11 Charge of an
R–C circuit

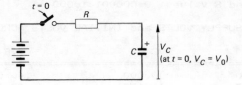

where E = d.c. supply voltage, volts
t = time in seconds
V_0 = capacitor voltage at $t = 0$
C = capacitance in farads
R = resistance in ohms

The evaluation of v_C is made within the FOR loop with time t being incremented between $t = 0$ and $t = 6CR$ in steps of $0 \cdot 4CR$. Finer or coarser increments may be obtained by altering the constants STEPSIZE and MAXNOSTEPS in the CONST section.

```
PROGRAM RCTRANSDC;

CONST    MAXNOSTEPS=15;
         STEPSIZE=0.4;

VAR    C,R,E,VO,VC,VR,T        :REAL;
       N                       :INTEGER;

BEGIN
 WRITE('ENTER C and R values: ');
 READ(C,R);
 WRITELN;
 WRITE('ENTER DC SUPPLY VOLTS and INITIAL VOLTS ACROSS C: ');
 READ(E,VO);
 WRITELN;
 WRITELN('    T            VC                   VR');
 WRITELN('================================================');
  T:=0.0;
   FOR N:= 0 TO MAXNOSTEPS DO
    BEGIN
      T:=T+C*R*STEPSIZE*N;
      VC:=E+(VO-E)*EXP(-T/(C*R));
      VR:=E-VC;
      WRITELN(T:10,VC:16:3,VR:20:3)
    END;
  WRITELN('================================================');
  WRITELN;
  WRITELN('C IS FULLY CHARGED AFTER ',4*C*R,' seconds approx.')
END.
```

Here is an example output from the program for the case:

$$C = 1\,\mu\text{F}, \quad R = 10\,\text{k}\Omega, \quad E = 140\,\text{V}, \quad V_0 = 0\,\text{V}$$

```
ENTER C and R values: 0.000001 10000

ENTER DC SUPPLY VOLTS and INITIAL VOLTS ACROSS C: 140 0

        T                   VC                      VR
=======================================================
   0.000E+00              0.000                 140.000
   4.000E-03             46.155                  93.845
   1.200E-02             97.833                  42.167
   2.400E-02            127.299                  12.700
   4.000E-02            137.436                   2.564
   6.000E-02            139.653                   0.347
   8.400E-02            139.969                   0.031
   1.120E-01            139.998                   0.002
   1.440E-01            140.000                   0.000
   1.800E-01            140.000                   0.000
   2.200E-01            140.000                   0.000
   2.640E-01            140.000                   0.000
   3.120E-01            140.000                   0.000
   3.640E-01            140.000                   0.000
   4.200E-01            140.000                   0.000
   4.800E-01            140.000                   0.000
=======================================================

C IS FULLY CHARGED AFTER  4.00000E-02 seconds approx.
```

4.7 The GOTO statement

The **GOTO** statement causes a direct transfer of control—a jump—from one part of a program to another. Most programs can be written without using GOTO by employing the control constructs considered in this Chapter. Indeed you should always aim to use these and avoid GOTO in order to preserve the structured nature of PASCAL. Unless your program can be made clearer—don't use GOTO statements. There are, however, a few advantages: if an input value is incorrect, GOTO can be used to exit immediately and simultaneously provide a warning message; GOTO can be used to exit from the middle of a WHILE or REPEAT loop, rather than from the "top" for WHILE or from the "bottom" for REPEAT loops. Although GOTO statements can be used to jump out of a conditional or repetitive statement (or procedure), they cannot be used to jump into one.

The GOTO statement has the simple form

GOTO (statement label);

where "statement label" is an unsigned integer which must lie in the range 0 to 9999. The GOTO statement transfers execution to the statement in the program which is prefixed by the "label-number" specified in the GOTO statment and a colon. For example,

. . . .

GOTO 100;

. . . .

. . . .

. . . .

100: WRITELN('$*$ $*$ Error in data entry $*$ $*$') END.

The GOTO 100 statement jumps execution to the statement "labeled" by 100: and in the above example this statement displays an error message and then the program is terminated by END. All statements between the GOTO and labeled statements are skipped.

Any label-number used in a program must be declared and all labels declared must be used. The label declaration section is positioned immediately before the CONST and VAR declaration sections. The syntax diagram for label declaration is given in fig. 4.12. If more than one label is required, each must have its own number (between 0 and 9999) and each number must be separated by a comma.

Fig. 4.12 Syntax diagram for **LABEL** declaration

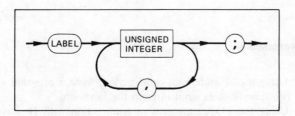

Example 4.8 This example illustrates the use of a GOTO statement in jumping out of a REPEAT loop if certain conditions occur.

The program simulates the testing of a batch of transistors. If a transistor gain is less than 20, we are notified; if more than 5 transistors in the first 10 fall below this specification, we are notified and the GOTO 100 statement immediately exits from the REPEAT loop to the statement prefixed by 100: which terminates the program. (*Note*: a label and a colon constitutes a statement even if no action follows the colon, other than END.).

```
PROGRAM EXGOTO1;

LABEL  100;

VAR  GAIN  :REAL;
     N,BAD :INTEGER;

BEGIN
 N:=0;BAD:=0;
   REPEAT
     N:=N+1;
     WRITE('INPUT GAIN: ');
     READ(GAIN);
     IF GAIN < 20 THEN
       BEGIN
         WRITE('TRANSISTOR NO:',N:3);
         WRITELN(' IS OUTSIDE SPEC.');
         BAD:=BAD+1
       END;
     IF (BAD=5) AND (N<=10) THEN
       BEGIN
         WRITELN('*** DISCARD BATCH ***');
         WRITELN('IN FIRST ',N:3,' TRANSISTORS');
         WRITELN(BAD:3, ' ARE BELOW SPECIFICATION');
         GOTO 100
       END;
   UNTIL
 N=20;
 WRITELN ('*** BATCH RESULTS ***');
 WRITELN(N-BAD:3,' WITHIN SPECIFICATION');
 WRITELN(BAD:3,' OUTSIDE SPECIFICATION');
 100:  END.
```

Exercises 4

4.1   a) Using the IF statement construction, write a program which displays the greater of two quantities entered in from the keyboard.

b) Using the FOR statement in conjunction with IF statements, write a program which will display the maximum and the minimum of 20 (say) values entered in from the keyboard.

4.2   Construct a program using the WHILE statement which continuously reads three voltages V_1, V_2, and V_3 and tests whether they are within the following ranges:

$$0 \leqslant V_1 \leqslant 1 \cdot 0, \quad 10 \leqslant V_2 \leqslant 12 \cdot 0, \quad 100 \leqslant V_3 \leqslant 120$$

If any of the voltages lie outside the specification the WHILE loop is to be exited and the fault value displayed.

4.3   Using the FOR statement write programs to
a) clear the screen of the VDU.
b) to produce a time delay.
c) to produce a table of values of x, e^{-x} and $(1-e^{-x})$ for $x = 0$ to $x = 4 \cdot 0$ in steps of $0 \cdot 2$

4.4 Construct a simulated thermostat control program which reads TEMPERATURE from the keyboard and
IF TEMPERATURE is less than or equal to 15 °C outputs
"switch on heater plus boost"
IF TEMPERATURE lies between 15 and 22 °C outputs
"heater on only"
IF TEMPERATURE is greater than 22 but less than 26 °C outputs
"switch off heater"
IF TEMPERATURE is greater than 26 °C outputs
"switch on fan".

4.5 Write a program in which numerical values are entered in from the keyboard (until a zero value is entered in as a terminator), counts the number of values, and determines the mean and root-mean-square (rms) value:

$$\text{Mean value} = \frac{\text{Sum of values}}{N}$$

$$\text{rms value} = \sqrt{\frac{\text{Sum of squares of values}}{N}}$$

$N =$ number of values

4.6 Modify the program of **4.5** to find a good approximation of the mean and rms values of
a) the half sinewave of fig. 4.13*a*
b) the sawtooth waveform of fig. 4.13*b*.

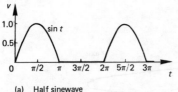

(a) Half sinewave

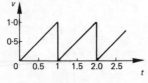

(b) Sawtooth waveform

Fig. 4.13

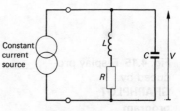

Fig. 4.14

4.7 The normalized frequency response of the parallel tuned circuit shown in fig. 4.14 is given approximately by

$$\frac{V}{V_0} = \frac{1}{\sqrt{(1 + 4Q^2 \delta^2)}}$$

where $Q = \dfrac{\omega_0 L}{R} = \dfrac{1}{R} \sqrt{\dfrac{L}{C}}$ is the circuit Q-factor

$$\delta = \frac{f - f_0}{f_0} \qquad f_0 = \text{resonant frequency}, \ f = \text{frequency}$$

$$\omega_0 = 2\pi f_0 \approx \frac{1}{\sqrt{(LC)}}$$

$V_0 =$ output voltage when $f = f_0$.

Construct a program which will tabulate the response of the circuit, V/V_0 versus δ, from $\delta = -0\cdot2$ in $0\cdot02$ steps to $\delta = +0\cdot02$ for $Q = 1, 10, 50, 100$.

4.8 The following program may be used for simple graph plotting of functions: it plots a function (identifier F in the program) suitably scaled in the y-direction against x for a given number of points. With reference to screen, the y-axis is horizontal and the x-axis runs vertically downwards (see fig. 4.15).

```
PROGRAM GRAPHPLOT1;

{A simple graph plotting program}

CONST   YZERO=40;
        NOOFPOINTS=30;
        YMULT=40;
        XSPACE=0.1;
        PI=3.14159;

VAR     X,Y,F     :REAL;
        N,M       :INTEGER;

BEGIN
 FOR N:=0 TO NOOFPOINTS DO
  BEGIN
   X:=XSPACE*N;
   F:=EXP(-X)*SIN(2*PI*X);
   Y:=YMULT*F+YZERO;
    FOR M:=1 TO ROUND(Y) DO
     WRITE(' ');
     WRITELN('*')
  END
END.
```

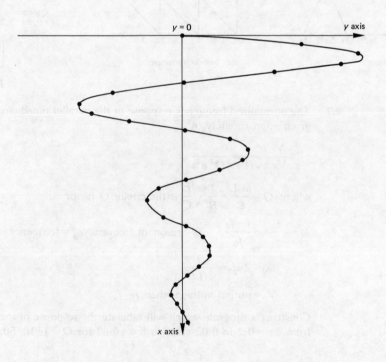

Fig. 4.15 Display produced by **GRAPHPLOT** program

The constants defined in the CONST section enable you

(i) To select the $y = 0$ position (i.e. x-axis); YZERO = 40 means, in the present program and with an 80-character width screen, that the x-axis runs vertically in the middle of the screen.

(ii) To scale, if necessary, your y values, e.g. YMULT = 40 multiplies the function value by 40.

(iii) To select the increment in x, e.g. XSPACE = 0·1 increments x from successive point plots by 0·1.

(iv) To plot for a given number of points (i.e. range of x), e.g. NOOFPOINTS = 30 will provide a plot of 30 + 1 points.

The above program plots $y = e^{-x} \sin 2\pi x$ over $2\frac{1}{2}$ cycles. Run the program. You should obtain the display shown in fig. 4.15b.

Amend the program to obtain a display of the following functions:

a) $y = 100(1 - e^{-x})$ for $x = 0$ to 4 in 0·2 steps

b) $y = 50 \cos^2 x$ for $x = 0$ to 2π in $0·1\pi$ steps.

4.9 A radio-frequency carrier, amplitude-modulated by a sinusoidal signal, can be represented by the expression (for double-sideband amplitude-modulation, DSB–AM):

$$v = A(1 + m \cos \omega_m t) \cos \omega_c t$$

where $m = \dfrac{\text{signal amplitude}}{\text{carrier amplitude}}$ (known as the modulation index)

A = carrier amplitude

$\omega_m = 2\pi f_m$, f_m = signal frequency

$\omega_c = 2\pi f_c$, f_c = carrier frequency

The frequency components of the DSB–AM wave are

Carrier: f_c, amplitude = A

Lower sideband: $f_c - f_m$, amplitude = $\frac{1}{2}mA$

Upper sideband: $f_c + f_m$, amplitude = $\frac{1}{2}mA$

The ratio of the power contained in the sidebands to the total power in the DSB–AM wave is

$$\frac{P_{SB}}{P_T} = \frac{m^2}{m^2 + 2}$$

Using the above information construct the following programs:

a) To accept carrier and signal amplitudes and frequencies from the keyboard and display m, sideband frequency components, percentage of total power contained in sidebands, and guard against an entry for which $m > 1$. (*Note*: the modulation index m must not exceed 1, otherwise severe distortion will result.)

b) To tabulate for $m = 0$ to 1 in steps of 0·1 the ratio P_{SB}/P_T.

c) To plot the envelope $A(1 + m \cos \omega_m t)$ for any value of m entered from the keyboard over 2 cycles of the signal frequency. [Hint: adapt program of **4.8**.]

Fig. 4.16 Square wave

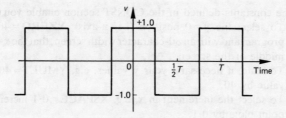

4.10 The Fourier series for the square wave of fig. 4.16 is given by

$$v = \frac{4}{\pi}(\cos \omega t - \tfrac{1}{3}\cos 3\omega t + \tfrac{1}{5}\cos 5\omega t - \tfrac{1}{7}\cos 7\omega t + \cdots)$$

where $\omega = 2\pi f = \dfrac{2\pi}{T}$

f = repetition frequency of wave, $\quad T = \dfrac{1}{f}$ its period,

i.e. the square wave can be considered as composed of a series of cosine waves of frequencies $f, 3f, 5f, 7f \ldots$

Construct a program which will tabulate, for 20 equally separated points over a complete cycle T of the square wave, approximations to the square wave by taking

the first (fundamental) harmonic only, i.e. $V_1 = \dfrac{4}{\pi}\cos \omega t$

the first and third, i.e. $V_3 = \dfrac{4}{\pi}(\cos \omega t - \tfrac{1}{3}\cos 3\omega t)$

the first, third, fifth, i.e. $V_5 = \dfrac{4}{\pi}(\cos \omega t - \tfrac{1}{3}\cos 3\omega t + \tfrac{1}{5}\cos 5\omega t)$

the harmonics up to and including $11f$.

5 Procedures and Functions

5.1 Introduction and summary

A **procedure** is essentially a self-contained subprogram designed to accomplish a given task and, once written, can be called into action to execute its task wherever and whenever this task is required in a program.

A **function** is a subprogram which is used to perform a specific calculation (or similar task) and returns a single result. Functions like procedures may be called into action wherever they may be required in a program.

The use of procedures and functions in program design has a number of very important advantages. They enable larger PASCAL programs to be broken down into a number of smaller and more manageable sections. They make program writing, reading, correction and understanding very much easier in all respects. In particular:

- Their use avoids duplication in a program; each procedure and function is written only once but may be "called", merely by writing its identifier, as many times as its action is required in the program.
- Larger and more complex programs are easier to develop—various steps in the solution can be developed, tested and improved individually and then subsequently incorporated in the complete program, each solution step being defined by a procedure.
- Programs written in terms of procedures are very much easier for other users (and yourself) to both read and understand.

In this chapter we consider how procedures and functions may be written, how information may be input and output and how they may be incorporated into PASCAL programs.

5.2 PASCAL programs: basic structure, order, global and local variables

The basic structure and the order in which a PASCAL program must be written, whether or not procedures and functions are included, is shown in fig. 5.1.

Following the main program heading we define any labels, constants, user-defined types and declare all variables that we wish to use throughout

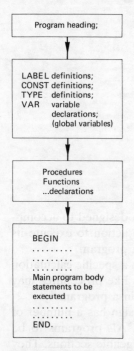

Fig. 5.1 Block diagram showing order in which various sections of a PASCAL program must be written

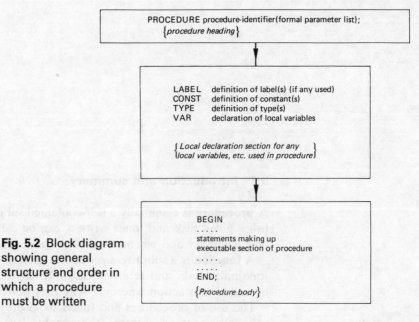

Fig. 5.2 Block diagram showing general structure and order in which a procedure must be written

the program. For this reason, any *identifier* defined in these sections is known as **global** to distinguish them from those variables, etc. which are declared and only valid *locally* within a procedure or a function. Global variables can be used anywhere in a program.

All procedures and functions are declared immediately after the global VAR section.

Procedure declarations have a similar structure and identical order to the main program as indicated in fig. 5.2. Each procedure commences with PROCEDURE and must be allocated its own identifier. Where information is to be passed to the procedure and/or from the procedure to the main program, a formal parameter list is defined immediately after the identifier (more about this in sections 5.4 and 5.5).

Following the procedure heading are the **local** definition/declaration sections. All variables defined here are for local use and are only valid within the procedure (or function) itself. They are local variables and are referred to as such. Unlike global variables, etc., local variables have no definition outside the procedure or function for which they have been defined.

5.3 Simple procedures

We consider first the simple type of procedure where no formal parameter list is used. The form of procedures of this type is shown in the syntax diagram of fig. 5.3.

Remember that the procedure must be allocated an identifier (its name, in the same way as for variables, etc.) and that this identifier follows immediately after PROCEDURE. Procedures are always declared im-

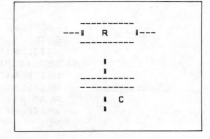

Fig. 5.3 Syntax diagram for a simple **PROCEDURE** (one without a formal parameter list)

mediately after the VAR section of the main program and are followed by the statement sections making up the main program.

A simple procedure is invoked, or called, to execute its task in the main program just by writing its identifier. For example, if we have a program containing procedures CALCZ and DISPLAYRESULT, they may be called into operation in the main program simply by the statements:

CALZ;
DISPLAYRESULT;

The following examples illustrate the basic ideas of how simple procedures are written, incorporated into the full program, and called.

```
PROGRAM DRAW1;

PROCEDURE DRAWR;
 BEGIN
  WRITELN;
  WRITELN('    -----------');
  WRITELN('---‖    R    I----');
  WRITELN('    -----------');
  WRITELN
 END;

PROCEDURE DRAWC;
BEGIN
  WRITELN('        I        ');
  WRITELN('        I        ');
  WRITELN('    ----------   ');
  WRITELN('    ----------   ');
  WRITELN('        I  C     ');
  WRITELN('        I        ')
END;

PROCEDURE CLEAR;
VAR N: INTEGER;
BEGIN
 FOR N:=1 TO 25 DO
   WRITELN
END;

BEGIN { Main program }
 CLEAR;
 DRAWR;
 DRAWC
END.
```

Example 5.1 This program contains three simple procedures:

DRAWR which creates a rather crude drawing of a resistor symbol.

DRAWC which "draws" a capacitor symbol.

CLEAR which "clears" the screen by writing 25 blank lines.

Fig. 5.4 Display obtained on running program DRAW1

Since we are not using any global variables, etc. in the main program, the procedures are declared immediately after the program heading. The statement section of the main program consists simply of a call to these procedures, i.e.

```
BEGIN
    CLEAR;      {call to procedure CLEAR to clear screen}
    DRAWR;      {call of DRAWR to draw R symbol}
    DRAWC;      {call of DRAWC to draw C symbol}
END.
```

Running the program gives the display shown in fig. 5.4—can you improve on the graphics for *R* and *C*?

Example 5.2 This program has a more practical use. It may be used to obtain design value for the resistors R_1 and R_2 in the T-network of fig. 5.5 which will provide a given attenuation when the T is inserted between a generator and load.

```
PROGRAM DESIGNPAD;

(* Design of a simple 3 element resistive pad, T-type *)

VAR  ADB,R              :REAL;
    {ADB ...variable for attenuation in decibels}

PROCEDURE DESIGN;
 VAR R1,R2,N            :REAL;
     { Declaration of local variables }
 BEGIN
 N:=EXP(ADB*LN(10.0)/20.0);
 {Above line converts dB to number ratio}
 R1:=R*(N-1.0)/(N+1.0);
 R2:=R*2.0*N/(N*N-1.0);
 WRITELN('*****************************************');
 WRITELN('FOR PAD ATTENUATION OF',ADB:6:1, 'dB');
 WRITELN('AND SYSTEM RESISTANCE', R:6:1,' Ohms');
 WRITELN('SERIES ELEMENTS,R1=',R1:8:2,' Ohms');
 WRITELN('SHUNT  ELEMENT, R2=',R2:8:2,' Ohms');
 WRITELN('*****************************************')
END;

PROCEDURE DATA;
BEGIN
 WRITELN('Enter attenuation required in decibels (dB)');
 READ(ADB);
 WRITELN(' ');
 WRITELN('Enter system (characteristic) resistance value');
 READ(R);
 WRITELN(' ')
END;

BEGIN  (* Main program *)
DATA;
DESIGN
END.
```

Fig. 5.5 Resistive pad and design equations: the T-network introduces an attenuation of 10 log N^2 decibels

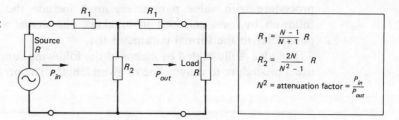

$$R_1 = \frac{N-1}{N+1}\ R$$

$$R_2 = \frac{2N}{N^2-1}\ R$$

$$N^2 = \text{attenuation factor} = \frac{P_{in}}{P_{out}}$$

DESIGN is the procedure which does the calculations and outputs the results.

DATA is the procedure used to input the attenuation required in dB (decibels) and the source and load value R (ohms).

On running the program to design, for example, a 3 dB, 600 Ω pad we obtain:

```
Enter attenuation required in decibels (dB)
3

Enter system (characteristic) resistance value
600

*******************************************
FOR PAD ATTENUATION OF    3.0dB
AND SYSTEM RESISTANCE 600.0 Ohms
SERIES ELEMENTS,R1=   102.60 Ohms
SHUNT  ELEMENT, R2= 1703.12 Ohms
*******************************************
```

5.4 Procedures with value parameters: passing information to a procedure

We can greatly enhance the versatility of a procedure by the inclusion of a list of parameters which will be used to supply *values* to be used in the execution of the procedure. This list, which is referred to as the **formal parameter list**, is declared immediately after the procedure identifier and in the way as shown in the syntax diagram of fig. 5.6. The formal parameter list defines variables (identifiers and type must both be given) which are to be used to input data (values) to the procedure, each time it is called. The variables in the list are therefore known as **value parameters**. A call to a

Fig. 5.6 Syntax diagram for a **PROCEDURE with VALUE PARAMETERS**

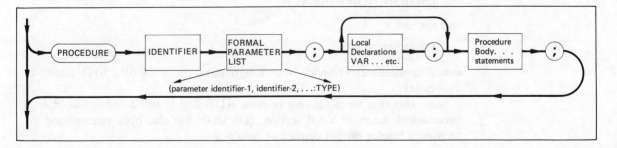

procedure with value parameters must include the procedure identifier, followed by, and enclosed in brackets, the **actual values** for each value parameter in the formal parameter list.

The idea is illustrated by means of the following simple example where we use a procedure to draw a line of given length. The procedure could take the form:

```
PROCEDURE PRINTALINE (LENGTH:INTEGER);
VAR    N  :INTEGER;
BEGIN
  FOR N:=1 TO LENGTH DO
    WRITE('–');
  WRITELN
END;
```

Following the procedure identifier PRINTALINE, we include the value parameter LENGTH and specify its type: INTEGER. We then declare the local variable N and use the FOR statement in the body of the procedure to draw the line. To call the procedure in the main program we can use, for example,

PRINTALINE(15);

This would cause the value parameter LENGTH to be assigned the value 15 and the procedure would display a line of LENGTH = 15 units.

As another example,

X:=15*2–5;   {where X is a global INTEGER variable}
PRINTALINE(X);

The **actual** parameter X is assigned the value of the expression $2*15-5 = 25$ by the first line and the call PRINTALINE(X) causes the value parameter LENGTH to be assigned the value of the actual parameter X (i.e. 25), so a line of 25 units is displayed.

Example 5.3 Here the procedure DESIGN of example 5.2 has been modified to include value parameters ADB and R (required attenuation and source/load resistance). Thus the procedure heading takes the form:

DESIGN(ADB,R:REAL);

and the call

DESIGN(20,50);

would immediately provide the design data for a 20 dB, 50 Ω resistive T-network.

Note also that we no longer require ADB and R to be defined as global variables in the main VAR section. (DESIGN has also been streamlined to provide a briefer output display of results.)

```
PROGRAM DESIGNPAD2;

(* Design of a simple 3 element resistive pad, T-type *)

VAR  COUNT :INTEGER; {...for use in Ex 2, see below}

PROCEDURE DESIGN(ADB,R:REAL);
VAR R1,R2,N            :REAL;
    { Declaration of local variables }
BEGIN
 N:=EXP(ADB*LN(10.0)/20.0);
 R1:=R*(N-1.0)/(N+1.0);
 R2:=R*2.0*N/(N*N-1.0);
 WRITE(' ATTENUATION =',ADB:6:1, 'dB');
 WRITELN('  SYSTEM RESISTANCE= ', R:6:1,' Ohms');
 WRITE('SERIES ELEMENTS,R1=',R1:8:2,' Ohms');
 WRITELN('  SHUNT  ELEMENT, R2=',R2:8:2,' Ohms');
 WRITELN
END;

BEGIN  {* Main program *}
 {Ex 1...this call gives design for
        ADB = 20 dB and R = 50 ohms}

    DESIGN(20,50);

 {Ex 2...use of FOR loop and call to DESIGN
        to produce table of design values
        for R fixed at 50 ohms (say)}

    FOR COUNT:=1 TO 10 DO
      DESIGN(COUNT,50);

END.
```

On running the program:

```
 ATTENUATION =  20.0dB  SYSTEM RESISTANCE=    50.0 Ohms
SERIES ELEMENTS,R1=   40.91 Ohms  SHUNT  ELEMENT, R2=   10.10 Ohms

 ATTENUATION =   1.0dB  SYSTEM RESISTANCE=    50.0 Ohms
SERIES ELEMENTS,R1=    2.88 Ohms  SHUNT  ELEMENT, R2= 433.34 Ohms

 ATTENUATION =   2.0dB  SYSTEM RESISTANCE=    50.0 Ohms
SERIES ELEMENTS,R1=    5.73 Ohms  SHUNT  ELEMENT, R2= 215.24 Ohms

 ATTENUATION =   3.0dB  SYSTEM RESISTANCE=    50.0 Ohms
SERIES ELEMENTS,R1=    8.55 Ohms  SHUNT  ELEMENT, R2= 141.93 Ohms

 ATTENUATION =   4.0dB  SYSTEM RESISTANCE=    50.0 Ohms
SERIES ELEMENTS,R1=   11.31 Ohms  SHUNT  ELEMENT, R2= 104.83 Ohms

 ATTENUATION =   5.0dB  SYSTEM RESISTANCE=    50.0 Ohms
SERIES ELEMENTS,R1=   14.01 Ohms  SHUNT  ELEMENT, R2=  82.24 Ohms

 ATTENUATION =   6.0dB  SYSTEM RESISTANCE=    50.0 Ohms
SERIES ELEMENTS,R1=   16.61 Ohms  SHUNT  ELEMENT, R2=  66.94 Ohms
```

5.5 Procedures with variable parameters: passing information back from a procedure

In the last section we considered the use of the formal parameter list to supply *input values* to a procedure. Formal parameters used in this way are known as *value parameters*. Value parameters are local to a procedure and are allocated memory space only while the procedure is being executed. On exit from the procedure all value parameter (and local variable) locations are effectively wiped clean. Value parameters are assigned the respective values of the given actual parameters only when the procedure is called. There is then no further interaction between the actual parameters and the value parameters even though the latter are locally processed within the procedure. Value parameters therefore cannot be used to pass back any results obtained within the procedure to the main program.

To overcome this difficulty PASCAL allows us to define, in addition to value parameters, **variable parameters** which allow results from a procedure to be passed back. Variable parameters are declared in the formal parameter list by preceding their identifiers by VAR as shown in the syntax diagram of fig. 5.7. For example,

PROCEDURE   EXAMPLE1(A,B,C:INTEGER;   VAR   X,Y:REAL);

declares the *value parameters* A, B, C (for inputting data) and the *variable parameters* X, Y (to be used to return results obtained within the procedure to the main program).

Fig. 5.7 Syntax diagram for a **PROCEDURE with formal parameter** with VALUE

Although no additional memory space is created during execution of the procedure, the results to be passed back and identified by the variable parameters are transferred to global variable locations and are therefore not "lost".

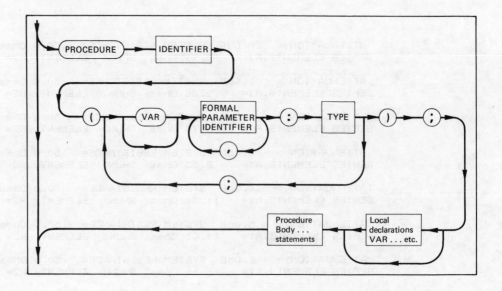

Example 5.4 This simple program uses the procedure FINDMAX to find the maximum value of three values entered from the keyboard. The *variable parameter* MAX in the formal parameter list provides the mechanism to pass back the maximum value found by the execution of the procedure to the main program.

Thus the procedure heading takes the form:

PROCEDURE FINDMAX(A,B,C:REAL; VAR MAX:REAL);

where A, B, C are *value parameters*, which are assigned respectively the values of the *actual parameters* (X, Y, Z · · · the global variables declared in the global VAR section) in the call

FINDMAX(X,Y,Z,HITEMP);

The *variable parameter* MAX is assigned to the global variable HITEMP in the above call and thus, after execution of FINDMAX, the value of MAX is placed in the storage location identified by the global variable HITEMP.

Likewise in the call

FINDMAX(X,Y,Z,RMAX);

the variable parameter MAX returns its value to the global variable RMAX. Try running the program and see that it works.

```
PROGRAM TEST2;
{ To illustate use of variable parameters to pass
  back results from a procedure to the main program }

VAR  X,Y,Z,HITEMP,RMAX :REAL;

PROCEDURE FINDMAX(A,B,C:REAL; VAR MAX:REAL);
 BEGIN
  IF A>=B THEN MAX:=A ELSE MAX:=B;
  IF C>MAX THEN MAX:=C
 END;

BEGIN
 WRITE('ENTER 3 TEMP VALUES '); READ(X,Y,Z);
         FINDMAX(X,Y,Z,HITEMP);
 WRITELN('Maximum temperature = ',HITEMP:6:1);
 WRITELN('Temperature limit   = ',1.5*HITEMP:6:1);

 WRITE('ENTER 3 RESISTOR VALUES '); READ(X,Y,Z);
         FINDMAX(X,Y,Z,RMAX);
 WRITE('Maximum Resistance   = ',RMAX:6:1)
END.
```

Fig. 5.8

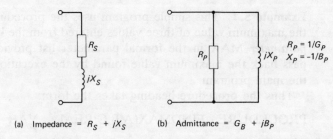

(a)   Impedance = $R_S + jX_S$ (b)   Admittance = $G_B + jB_P$

Example 5.5   The following program uses the procedure ADMITTANCE to determine the admittance of a network given its impedance, i.e. referring to fig. 5.8a,

Impedance $Z = R_S + jX_S$

Admittance $Y = \dfrac{1}{Z} = \dfrac{1}{R_S + jX_S}$

$$= \frac{1}{R_S + jX_S} \times \frac{R_S - jX_S}{R_S - jX_s}$$

$$= \frac{R_S}{R_S{}^2 + X_S{}^2} - j\frac{X_S}{R_S{}^2 + X_S{}^2}$$

$$\equiv G_P + jB_P$$

The procedure does the above calculations and passes back the results, which we then use in the main program to determine the equivalent parallel elements R_P and X_P of the network (see fig. 5.8b).

```
PROGRAM CIRCUITCALC2;

{ To determine admittance and equivalent parallel
  elements of a component from its impedance }

VAR  RS,XS,GP,BP    :REAL;

PROCEDURE ADMITTANCE(R,X:REAL; VAR G,B:REAL);
BEGIN
  G:= R/(R*R+X*X);
  B:=-X/(R*R+X*X);
  WRITELN('Admittance = ',G:6:4,' +j ',B:6:4)
END;

BEGIN   {Main program}
  WRITE('Enter series resistance and reactance: ');
  READ(RS,XS);
  ADMITTANCE(RS,XS,GP,BP);
  WRITELN('Equivalent parallel components are:');
  WRITELN('RP = ',1/GP:6:1,'   XP = ',-1/BP:6:1)
END.
```

On running the program and entering RS = 100, XS = 50 we obtain the display:

```
Enter series resistance and reactance: 100 50
Admittance = 0.0080 +j -4.00000E-03
Equivalent parallel components are:
RP =  125.0  XP =   250.0
```

5.6 Using global variables with procedures to input and output information

Global variables as distinct from local variables and parameters remain valid throughout the program. They can therefore be used—but with care—in association with procedures:

1 *To pass back results and hold values*
The procedure itself can be used to assign values to global variables. There must, of course, be no local variables within the procedure with the same identifiers as used for the global variables.

There are, however, certain drawbacks to changing a global value in a procedure known collectively as **side effects**. One undesirable side effect is an accidental change in global variable value within a procedure, which in practice may be very difficult to detect. The use of variable parameters is normally preferred to passing back data as this overcomes the side-effect problem.

2 *To input data to a procedure*
Global variables are often used to read data into a procedure. For example, using READ statements global variables may be assigned values which are then used to initialize the value parameters prior to a procedure call (see examples 5.4 and 5.5).

Examples 5.6 This program contains procedures for the addition, subtraction, multiplication and division of complex numbers, i.e.

ADD   performs addition of two complex numbers
    $X_1 + jY_1$ and $X_2 + jY_2$
SUBTRACT   performs the subtraction: $(X_1 + jY_1) - (X_2 + jY_2)$
MULTIPLY   performs the multiplication: $(X_1 + jY_1)(X_2 + jY_2)$

DIVIDE   performs the division: $\dfrac{X_1 + jY_1}{X_2 + jY_2}$

The procedure POLAR converts from cartesian to polar form, i.e. from $X + jY$ to $R\underline{/\theta}$.

The procedures are used in the main program to find the transfer function:

$$\frac{V_o}{V_i} = \frac{Z_2}{Z_1 + Z_2}$$

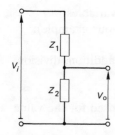

Fig. 5.9

of the network shown in fig. 5.9.

```
PROGRAM COMPLEXNO1;

(* Program contains procedures for manipulation
   of complex numbers and applies these to
   the calculation of a simple transfer function *)

VAR X1,X2,Y1,Y2,X,Y,Z,ANGLE  :REAL;

PROCEDURE ADD(X1,Y1,X2,Y2:REAL);
BEGIN
  X:=X1+X2;Y:=Y1+Y2
END;

PROCEDURE SUBTRACT(X1,Y1,X2,Y2:REAL);
BEGIN
  X:=X1-X2;Y:=Y1-Y2
END;

PROCEDURE MULTIPLY(X1,Y1,X2,Y2:REAL);
BEGIN
  X:=X1*X2-Y1*Y2;Y:=Y1*X2+X1*Y2
END;

PROCEDURE DIVIDE(X1,Y1,X2,Y2:REAL);
VAR D:REAL;
BEGIN
  D:=X2*X2+Y2*Y2;
  X:=(X1*X2+Y1*Y2)/D;Y:=(Y1*X2-X1*Y2)/D
END;

PROCEDURE POLAR(X,Y:REAL);
BEGIN
  Z:=SQRT(X*X+Y*Y);ANGLE:=ARCTAN(Y/X)
END;

BEGIN (* MAIN PROGRAM *)
  WRITE('Enter resistive and reactive components of Z1: ');
  READ(X1,Y1);
  WRITE('Enter resistive and reactive components of Z2: ');
  READ(X2,Y2);
  ADD(X1,Y1,X2,Y2); { gives Z1+Z2 }
  DIVIDE(X2,Y2,X,Y);{ gives Z2/(Z1+Z2) }
  WRITELN('TRANSFER FUNCTION =',X:8:4,' +j(',Y:8:4,' )');
  POLAR(X,Y);
  WRITELN('IN POLAR FORM: MAGNITUDE = ',Z:8:4);
  WRITELN('              PHASE ANGLE = ',ANGLE*180/3.14159:6:1)
END.
```

Note that the program contains declarations of global variables:

X1, Y1 for one complex number (or impedance Z1 in our example);

X2, Y2 for the second complex number;

X, Y for the real and imaginary parts of the results for addition, division, etc;

Z for the magnitude and ANGLE for the polar results.

Within the procedures themselves the same notation is used for the value parameter identifiers.

In the main program data is read in using the READ statement and then the following procedure calls are used:

ADD to find $Z_1 + Z_2$, $Z_1 + Z_2 = X + jY$

DIVIDE(X1,Y2,X,Y) to find $\dfrac{Z_2}{Z_1 + Z_2} = \dfrac{X_2 + jY_2}{X + jY} \rightarrow X + jY$

POLAR(X,Y) to convert the transfer function $X + jY$ to polar form

We use global variables to initialize the value parameter and also to pass back or, rather, pass on results from one procedure to the next.

Try running the program. Here is the output obtained for the case

$$Z_1 = 100 + j15 \quad \text{and} \quad Z_2 = 40 - j12$$

```
Enter resistive and reactive components of Z1: 100 15
Enter resistive and reactive components of Z2: 40 -12
TRANSFER FUNCTION =  0.2837 +j( -0.0918 )
IN POLAR FORM: MAGNITUDE =    0.2982
               PHASE ANGLE =  -17.9
```

5.7 User-defined functions

We have already met many of the standard functions available in PASCAL, e.g. SIN(X), SQRT(X), CHR(X), in Chapter 3. We now consider how we can define and subsequently use functions. A user-defined function is very similar in structure to a procedure. The basic difference is that a function returns a single result when called, whereas a procedure may return several results or even none at all but instead carry out a number of tasks.

The syntax diagram for a function is shown in fig. 5.10. This is similar to that of a procedure except that the result type must be included (and, obviously, the formal parameter list contains only value parameters). The formal parameter list contains the variables (identifiers and type) which form the input data to the function. The function is called in the main program by writing its identifier and assigning values to the formal (value) parameters. The following example illustrates how functions may be written and called.

Fig. 5.10 Syntax diagram for a **FUNCTION**

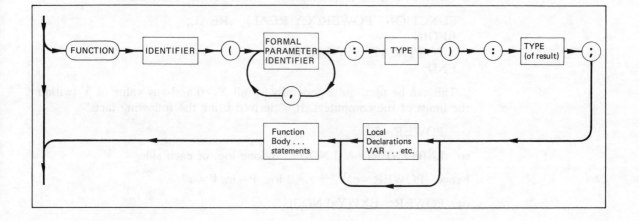

Example 5.7 The following function CELSIUS converts degrees Fahrenheit to Celsius:

```
FUNCTION  CELSIUS(F:REAL)  :REAL;
BEGIN
  CELSIUS:=(F−32)*5/9
END;
```

and the call

```
CELSIUS(128·6);
```

in the main program would return $(128·6 − 32) * 5/9 = 53·6667$.

Example 5.8 Unfortunately, PASCAL provides no direct means of computing powers or roots of numbers. The following user-defined functions do the job.

1 *Function to calculate A^N where A and N are positive whole numbers.*

```
FUNCTION  POWER(A,N:INTEGER)  :INTEGER;
VAR    I  :INTEGER;
BEGIN
  I:=0;POWER:=1;
    WHILE I<N DO
      BEGIN
        POWER:=POWER*A;
        I:=I+1
      END
END;
```

Try incorporating this in a program but remember computers have a limited range for INTEGERS so be careful not to exceed the MAXINT value.

2 *General method to calculate X^Y where X and Y are REAL*

```
FUNCTION  POWER(X,Y:REAL)  :REAL;
BEGIN
  POWER:=EXP(Y*LN(X))
END;
```

This can be used quite generally for all $X > 0$ and any value of Y (within the limits of the computer). It is derived using the following facts:

$$POWER = X^Y$$

so $LN(POWER) = Y\ LN(X)$ taking $\log_e$ of each side

hence $POWER = e^{Y LN(X)}$ if $\log_e P = m,\ P = e^m$

or $POWER := EXP(Y*LN(X))$

Example 5.9 The capacitance C_D of a certain varactor diode varies with reverse bias voltage V_B according to:

$$C_D = \frac{C_0}{(1+V_B/V_0)^G}\,\text{pF}$$

where $C_0 = 10\,\text{pF}$, $V_0 = 0\cdot52\,\text{V}$, $G = 0\cdot45$.

Write a program to draw up a table of C_D values as V_B varies in 1 volt steps from 0 to 10 V.

Here is a possible solution using the function C_D to evaluate the capacitance.

```
PROGRAM VARACTORC;

VAR  VB :INTEGER;

FUNCTION CD(VB:REAL):REAL;
CONST  CO=10; VO=0.52; G=0.45;
 BEGIN
  CD:=CO*EXP(-G*LN(1+VB/VO))
 END;

BEGIN
 WRITELN('BIAS VOLTAGE':20,'CAPACITANCE, CD':20);
 WRITELN('volts':16,'pico-farads':22);
 WRITELN('          -------------------------------------');
  FOR VB:=1 TO 10 DO
    WRITELN( VB:15,CD(VB):20:3)
END.
```

On running the program, the following table of results is displayed:

```
    BIAS VOLTAGE        CAPACITANCE, CD
       volts               pico-farads
    ------------------------------------
            1                   6.171
            2                   4.915
            3                   4.229
            4                   3.779
            5                   3.454
            6                   3.205
            7                   3.005
            8                   2.841
            9                   2.703
           10                   2.584
```

Fig. 5.11 Circuit for a voltage-controlled oscillator

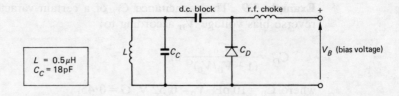

L = 0.5µH
C_C = 18pF

Example 5.10   Figure 5.11 shows the equivalent circuit of a varactor-tuned voltage-controlled oscillator. The oscillation frequency of the circuit is

$$f = \frac{1000}{2\pi\sqrt{[L(C+C_D)]}} \text{ MHz} \qquad L \text{ in micro-henries, } C, C_D \text{ in pico-farads}$$

where C_D is the varactor diode capacitance. Assuming that C_D is given by the same formula as in Example 5.9 write a program that will display a table of oscillator frequency versus reverse bias voltage V_B as V_B is varied between 0 and 10 V in 0·5 V steps.

Here is a possible solution:

```
PROGRAM VCO;

{VARACTOR CONTROLLED OSCILLATOR}

CONST     L=0.5; CC=18; PI=3.14159;
          CO=10; VO=0.52; G=0.45;

VAR       VB,F      :REAL;
          N         :INTEGER;

FUNCTION CD(VB:REAL):REAL;
  BEGIN
    CD:=CO*EXP(-G*LN(1+VB/VO))
  END;


BEGIN
  WRITELN(' BIAS VOLTAGE            FREQUENCY ');
  WRITELN('      volts               MHz      ');
  WRITELN('=================================');
  VB:=0.0;
  FOR N:=0 TO 20 DO
    BEGIN
      F:=1000/(2*PI*SQRT(L*(CD(VB)+CC)));
      WRITELN(VB:10:1,F:20:3);
      VB:=VB+0.5
    END
END.
```

and the results obtained on running the program:

BIAS VOLTAGE volts	FREQUENCY MHz
0.0	42.536
0.5	44.673
1.0	45.781
1.5	46.500
2.0	47.019
2.5	47.418
3.0	47.739
3.5	48.005
4.0	48.230
4.5	48.424
5.0	48.594
5.5	48.744
6.0	48.879
6.5	49.000
7.0	49.110
7.5	49.211
8.0	49.303
8.5	49.388
9.0	49.468
9.5	49.541
10.0	49.610

Exercises 5

5.1 With reference to PASCAL programs distinguish between the following:
a) procedures and functions
b) local and global variables
c) value and variable parameters
d) formal and actual parameters.

5.2 Write functions RSERIES and RPARALLEL which will determine, respectively, the total resistance of 3 resistors connected in series and 3 resistors connected in parallel, as shown in fig. 5.12*a* and *b*.

Use these two functions in a program which calculates the total resistance of the circuit shown in fig. 5.12*c*.

Fig. 5.12

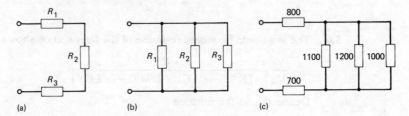

(a)　　　　　　(b)　　　　　　(c)

5.3 Write a procedure which utilizes the result:

$$x = \frac{-b \pm \sqrt{(b^2 - 4ac)}}{2a}$$

and returns the two solutions for the general quadratic equation:

$$ax^2 + bx + c = 0$$

Incorporate this procedure in a program which allows you to enter the coefficients a, b and c of a quadratic equation from the keyboard and displays the two solutions. Run your program to obtain the solutions for the following equations:

a) $x^2 - 2x - 6 = 0$ b) $3x^2 + 7x - 24 = 0$ c) $12x^2 + 5x + 3 = 0$

5.4 Write functions which will return the following values:
a) the average value of 3 quantities
b) the maximum of 3 quantities
c) the minimum of 3 quantities
d) $\tan(A)$, remember $\tan A = \sin A/\cos A$.

5.5 Write the following procedures:
a) To add two complex impedances

$$Z_1 = R_1 + jX_1 \quad \text{and} \quad Z_2 = R_2 + jX_2$$

in series (see fig. 5.13a).
b) to determine the admittance of Z_1 and Z_2 in parallel (see fig. 5.13b).

Hence or otherwise construct a program to determine the input impedance of the circuits shown in fig. 5.13c and d for the impedance values shown.

Fig. 5.13

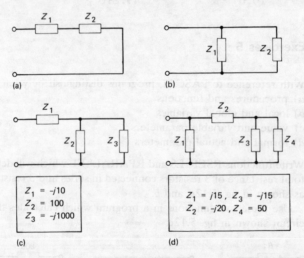

(a)

(b)

(c)

$Z_1 = -j10$
$Z_2 = 100$
$Z_3 = -j1000$

(d)

$Z_1 = j15$, $Z_3 = -j15$
$Z_2 = -j20$, $Z_4 = 50$

5.6 The amplitude frequence response of the filter section shown in fig. 5.14 is given by

$$\frac{V_0}{V_{in}} = \frac{R}{\sqrt{[R^2(1 - \omega^2 LC)^2 + L^2\omega^2(2 - \omega^2 LC)^2]}}$$

Define this as the function

V0(R,L,C,F :REAL) :REAL

where $F = \text{frequency} = \omega/2\pi$. Using this function construct a program that tabulates the response for the case $L = 0 \cdot 75$ mH, $C = 0 \cdot 5\ \mu$F, $R = 50\ \Omega$ over the frequency range 1 kHz to 20 kHz in 1 kHz steps.

5.7 The trapezoidal rule for finding the area under a curve (see fig. 5.15) is given by

$$\text{Area} = h[\tfrac{1}{2}(y_1 + y_N) + y_2 + y_3 + \cdots + y_N]$$

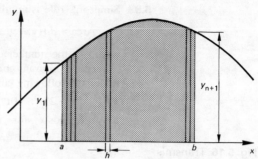

Fig. 5.14

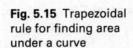

Fig. 5.15 Trapezoidal
rule for finding area
under a curve

where h = width of the individual strips,
        n = number of strips,

so $h = \dfrac{b-a}{n}$, $a = x$ lower-limit, $b = x$ upper-limit

        $y, y_2 \cdots y_{n+1}$ = ordinates (y-values) of individual strips.

The following program includes the procedure TRAPRULE to calculate the area
under the curve $y = 3x^2 + 5x + 6$ for two cases $n = 10$ and $n = 100$ strips between the
limits $a = 1$ and $b = 12$. Try running the program and compare your results with the
accurate value obtained by integration:

```
PROGRAM INTEGRATE;

VAR AREAn                 :REAL;

PROCEDURE TRAPRULE(A,B:REAL;N:INTEGER;VAR AREA:REAL);
VAR H,SUMOFY :REAL;
    M            :INTEGER;
FUNCTION Y(X:REAL):REAL;
BEGIN
 Y:=3*X*X+5*X+6
END;
BEGIN {NUMERICAL INTEGRATION}
 H:=(B-A)/N;SUMOFY:=0;
  FOR M:=1 TO N-1 DO
   SUMOFY:=SUMOFY+Y(A+M*H);
   SUMOFY:=SUMOFY+0.5*(Y(A)+Y(B));
 AREA:=SUMOFY*H
END;


BEGIN { Main program }
  TRAPRULE(1,12,10,AREAn);
  WRITELN('AREA for n=10 strip approx. is :');
  WRITELN(AREAn:15:4);
  TRAPRULE(1,12,100,AREAn);
  WRITELN('AREA for n=100 strip approx. is :');
  WRITELN(AREAn:15:4)
END.
```

Area under a curve is the definite integral $\displaystyle\int_a^b y\,dx$. Adapt the procedure TRAP-
RULE to evaluate the area under the curve for the following functions:
a) $y = \sin x$ for $x = 0$ to $x = \pi$
b) $y = e^{-x}$ for $x = 0$ to $x = 5$.

5.8   Simpson's rule is another numerical method for finding areas under curves:

$$\text{Area} = \tfrac{1}{3}h[(y_1 + y_{n+1}) + 2(y_3 + y_5 + \cdots + y_{n-1}) + 4(y_2 + y_4 + \cdots + y_n)]$$

where n (the number of strips) is an even number.

Write a program containing a procedure SIMPSONRULE to calculate the area under the following curves:

a) $v = 2t^3 - 21t^2 + 60t$ for $t = 0$ to $t = 6$

b) $y = e^{-x}\cos x$ for $x = 0$ to 5.

Fig. 5.16 Transmission line and equivalent circuit of a short length in terms of distributed line parameters

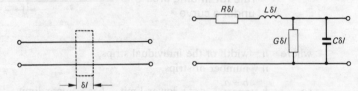

R = series resistance per unit length of line, Ω/m
L = series inductance per unit length of line, H/m
G = shunt conductance between line conductors per unit length, S/m
C = shunt capacitance between line conductor per unit length, F/m

5.9   The characteristic impedance Z_0, attenuation coefficient α and phase constant β of a transmission line are given in terms of its distributed parameters (see fig. 5.16) by the following formulae:

$$Z_0 = \sqrt[4]{\left(\frac{R^2 + \omega^2 L^2}{G^2 + \omega^2 C^2}\right)}\,\Omega$$

$$\alpha = \frac{1}{\sqrt{2}}\sqrt{[(RG - \omega^2 LC) + \sqrt{\{(R^2 + \omega^2 L^2)(G^2 + \omega^2 C^2)\}}]}\ \text{nepers/m}$$

$$\beta = \frac{1}{\sqrt{2}}\sqrt{[(\omega^2 LC - RG) + \sqrt{\{(R^2 + \omega^2 L^2)(G^2 + \omega^2 C^2)\}}]}\ \text{radians/m}$$

Write functions to determine the above three line parameters with the distributed parameters R, L, C, G and frequency $(f = \omega/2\pi)$ as value parameters. Hence construct a program which tabulates Z_0, α and wavelength λ $(\lambda = \beta/2\pi)$ for $f = 10$ Hz, 100 Hz, 1 kHz, 100 kHz and 1 MHz.

5.10   Utilizing the functions developed in **5.9** determine the characteristic impedance and attenuation coefficient of a coaxial cable used in multichannel telephony at $f = 3$ MHz given that line parameters are:

$R = 65\ \Omega$/km, $L = 0\cdot27$ mH/km, $C = 0\cdot04\ \mu$F/km
$G = 0\cdot002$ S/km

Calculate also the minimum gain in decibels of a repeater amplifier inserted at the mid-point of a 20 km section of line so as to achieve an overall system loss of less than 40 dB between sending and receiving ends.

Note: 1 neper = $8\cdot686$ dB.

6 The Use of Arrays and Files

6.1 Introduction and summary

So far we have used a single unique identifier for each variable employed in a program. Many programming tasks are concerned with the collection and subsequent processing of large amounts of related data. For such cases, the use of arrays to store and process data in a general and systematic way offers considerable scope to our programming ability.

In this chapter we begin by explaining what an array is, its forms and how it is declared. We then consider several applications of their use in programs, including sorting data and solving equations.

We then consider the use of files, introducing the topic by explaining the concept of a file in PASCAL and its advantage in providing a means to store input and output data on a permanent basis. We then explain how files are declared; how they may accessed for reading data (i.e. inputting data from an external file to a program); how they may be created and written to (i.e. opening a file so that data can be output and stored in that file). Finally some practical examples of using files in programs are given.

6.2 Arrays: their form and declaration

Fig. 6.1 gives a pictorial representation which should help you in understanding the concept of an array for storing data.

Each **array** used in a program is given its own general identifier, just as we assign names or identifiers to variables. Thus we can define a collective name for an associated group of variables rather than a whole list of different identifiers. Every element in an array represents a variable storage location in its own right. Elements of a given array must all be of the same type. We can easily refer to an individual element by using the array identifier plus number subscript(s), the number(s) specified within the [ ] brackets in fig. 6.1.

A syntax diagram for the shorthand form of array declaration is shown in fig. 6.2. Note that the number of elements in the array is defined by the constants (subscripts) I and J; I and J are normally of type INTEGER but any ordinal type can be used, e.g. CHAR or user-defined enumerated type. REAL subscripts obviously cannot be used.

Fig. 6.1 Pictorial representation of arrays: each box in the array can be considered as a storage location for the value of a variable

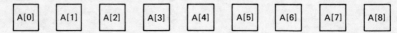

(a) Example of a one-dimensional array

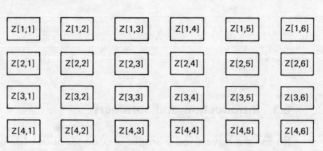

(b) Example of a two-dimensional array

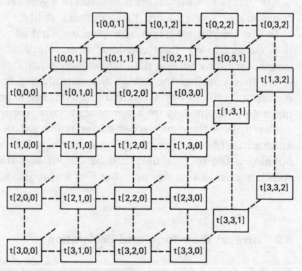

(c) Example of a three-dimensional array

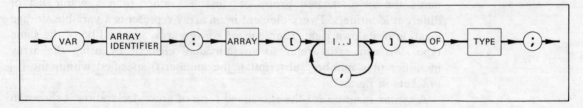

Fig. 6.2 A syntax diagram for the **declaration** of an ARRAY

Examples of array declaration

1 One-dimensional arrays

VAR   A: ARRAY[0..8] OF REAL;

         ↑   ↑ ↑   ↑

      array   subscripts determining type of

      identifier number of elements element

               in the array

The above declaration would cause an array of 9 elements (see fig. 6.1*a*):

A[0], A[1], A[1], A[2], A[3], . . . A[7], A[8]

of type REAL to be created, each individual element identifier being denoted by A[N] where $N = 0, 1, 2 \cdots 8$.

2 Two-dimensional arrays

VAR   Z[1..4, 1..6]: ARRAY OF REAL;

creates a two-dimensional array, identifier Z, consisting of a total $4 \times 6 = 24$ elements (see fig. 6.1*b*) of type REAL:

Z[1, 1] Z[1, 2] Z[1, 3] Z[1, 4] Z[1, 5] Z[1, 6]

Z[2, 1] Z[2, 2] Z[2, 3] Z[2, 4] Z[2, 5] Z[2, 6]

Z[3, 1] Z[3, 2] Z[3, 3] Z[3, 4] Z[3, 5] Z[3, 6]

Z[4, 1] Z[4, 2] Z[4, 3] Z[4, 4] Z[4, 5] Z[4, 6]

3 Multi-dimensional arrays

Arrays with more than two pairs of subscripts are defined in the same manner. There is no theoretical limit to the dimensions (size) of an array but in practice there will always be storage limitations dependent on the available memory of the computer.

Fig. 6.1*c* illustrates a 3-dimensional array:

VAR   t[0..3, 0..3, 0..2]: ARRAY OF REAL;

This declaration declares a $4 \times 4 \times 3$ element array of type REAL and identifier t.

So far we have used the short-hand form of declaration:

VAR   identifier: ARRAY[constant..constant] OF element type;

The more formal definition gives the array-type in the TYPE section and the array identifier in the VAR section:

TYPE   array-type = ARRAY[subscripts] OF element type;

VAR   identifier: array type

So, for example, if we wish to declare the following arrays:

VOLTAGE,CURRENT   one-dimensional arrays of 6 elements

ZMATRIX,YMATRIX   two-dimensional arrays of 6×6 elements

we could use the full declaration form of:

TYPE   SUBSCRIPTS = 1..6;

COLVECTOR = ARRAY[SUBSCRIPTS] OF REAL;

MATRIX = ARRAY[SUBSCRIPTS, SUBSCRIPTS] OF REAL;

VAR   VOLTAGE,CURRENT: COLVECTOR;

ZMATRIX,YMATRIX: MATRIX;

or in the short-hand form:

VAR   VOLTAGE,CURRENT=ARRAY[1..6] OF REAL;

ZMATRIX,YMATRIX=ARRAY[1..6,1..6] OF REAL;

6.3   The application of arrays in programs

The use of arrays saves a formidable amount of time and effort in preparing programs and allows the use of more general methods for inputting, processing and outputting large volumes of related information. Before considering some practical examples, we first summarize some important points:

1 An array must be declared in the VAR or TYPE and VAR sections at the front end of the program.

2 The array declaration assigns a single identifier to the group of related variables; the subscripts I..J specify the number of elements in the array; finally the type of the array elements must also be stated.

3 Array elements may be accessed by specifying the array identifier followed and enclosed in square brackets the "positional" subscript values, e.g.

WRITE(Z[1, 1]);

would display the value of element $Z[1, 1]$ of the array Z.

Example 6.1: Reading in values to array elements

This program illustrates how a number of items can be read in to an array and how to output this information. We use the FOR loop construct for both reading in and reading out.

FOR statements go hand in hand with processing arrays. They are ideally suited for stepping through the positional subscript values so that all elements making up the array can be simply and directly referenced.

In the program, **READIN** is the procedure for inputting data for the array from the keyboard; **READOUT** is a procedure for displaying the data stored in the array elements.

```
PROGRAM ARRAYTEST1;

{ To illustrate use of FOR statement to
  read in data to an array and also to
  output the data }

VAR  ITEM :ARRAY[1..10] OF INTEGER;

PROCEDURE READIN;
VAR N: INTEGER;
BEGIN
 FOR N:=1 TO 10 DO
  BEGIN
   WRITE('Enter item ',N,': ');
   READ(ITEM[N])
  END
END;

PROCEDURE READOUT;
VAR N: INTEGER;
 BEGIN
  FOR N:=1 TO 10 DO
  WRITELN('ITEM[',N,'] =',ITEM[N]:6)
 END;

BEGIN { Main program }
 READIN;
 READOUT
END.
```

Example 6.2: Reading in, simple processing, and outputting results

This program is a bit more ambitious. Two arrays are first declared to hold
numbers and unit prices of (say) stock items. The procedure READIN
allows us to enter in data for 10 items. Procedure DISPLAYRESULTS
displays the data and also works out sub-totals and total value of all stock.

```
PROGRAM ARRAYTEST2;

{ To read in data (number and price)
  To display table of data:
     no.  price  sub-total
  and total value of all items }

VAR  NUMBER: ARRAY[1..10] OF INTEGER;
     PRICE : ARRAY[1..10] OF REAL;

PROCEDURE READIN;
VAR N: INTEGER;
BEGIN
 FOR N:= 1 TO 10 DO
  BEGIN
   WRITE('Enter no. of items and unit price: ');
   READ(NUMBER[N],PRICE[N])
  END
END;
```

```
PROCEDURE DISPLAYRESULTS;
VAR N      :INTEGER;
    SUM    :REAL;
BEGIN
 SUM:=0;
 WRITELN(' No. of items':15,'Unit price':15,'Sub-total':15);
 WRITELN('-----------------------------------------');
  FOR N:=1 TO 10 DO
   BEGIN
    WRITE(NUMBER[N]:10,PRICE[N]:15:2);
    WRITELN(NUMBER[N]*PRICE[N]:15:2);
    SUM:=SUM+NUMBER[N]*PRICE[N]
   END;
 WRITELN('-----------------------------------------');
 WRITELN('TOTAL = ',SUM:10:2)
END;

BEGIN { Main program }
 READIN;
 DISPLAYRESULTS
END.
```

Here is an example of the output obtained on running the program (at least that part due to DISPLAYRESULTS):

```
Enter no. of items and unit price: 125 56
Enter no. of items and unit price: 34 12.78
Enter no. of items and unit price: 129 1.99
Enter no. of items and unit price: 45 7.90
Enter no. of items and unit price: 548 0.60
Enter no. of items and unit price: 12 99.85
Enter no. of items and unit price: 185 1.30
Enter no. of items and unit price: 77 0.12
Enter no. of items and unit price: 876 9.34
Enter no. of items and unit price: 111 55.98
     No. of items      Unit price       Sub-total
-----------------------------------------
          125            56.00          7000.00
           34            12.78           434.52
          129             1.99           256.71
           45             7.90           355.50
          548             0.60           328.80
           12            99.85          1198.20
          185             1.30           240.50
           77             0.12             9.24
          876             9.34          8181.84
          111            55.98          6213.78
-----------------------------------------
TOTAL =    24219.07
```

Example 6.3: Binary-decimal conversions
This program contains two procedures:

BINARYTODEC   to convert binary to decimal

DECTOBINARY   to convert decimal to binary

We use the array X[0..15] to store up to 16 binary 1s and 0s.

```
PROGRAM BINARYCONVERSIONS;

{BINARY TO DECIMAL CONVERSION}

VAR SELECT:INTEGER;

PROCEDURE BINARYTODEC;
VAR     N,NOOFBITS,BIT,DECNO :INTEGER;
        X:ARRAY[0..15] OF INTEGER;
BEGIN
WRITE('ENTER NO. OF BITS IN NUMBER');
READ(NOOFBITS);
WRITELN('ENTER BITS BEGINNING WITH MOST');
WRITELN('SIGNIFICANT BIT');
DECNO:=0;
 FOR N:=NOOFBITS-1 DOWNTO 0 DO
   BEGIN
     READ(BIT);
     X[N]:=ROUND(BIT*EXP(N*LN(2)));
     DECNO:=DECNO+X[N]
   END;
WRITELN('DECIMAL NO. =',DECNO )
END;


PROCEDURE DECTOBINARY;
VAR NO,P,DECNO,N  :INTEGER;
    X:ARRAY[0..15] OF INTEGER;
BEGIN
WRITELN('ENTER DECIMAL NO.');
READ(DECNO);
NO:=DECNO DIV 2;X[1]:=DECNO MOD 2;
N:=2;
 REPEAT
 X[N]:=NO MOD 2;
 NO:=NO DIV 2;
 N:=N+1
 UNTIL NO=0;
FOR P:=N-1 DOWNTO 1 DO
WRITE(X[P]:2);
WRITELN('..... THE BINARY NO')
END;

BEGIN { MAIN PROGRAM }
 WRITELN(' Enter 1 for binary to decimal conversion');
 WRITELN(' OR    2 for vice-versa');
 WRITELN('REMEMBER do not exceed MAXINT or 65535');
 READ(SELECT);
  IF SELECT=1 THEN  BINARYTODEC;
  IF SELECT=2 THEN  DECTOBINARY
END.
```

Example 6.4: Frequency distribution and bar chart using arrays

This example illustrates how we might construct a program to divide a list of values into a number of sub-ranges to determine the frequency distribution. The program also plots the results in the form of a bar chart. For example, suppose we had a batch of nominally 100 Ω resistors and we wished to count the number in the following bands: below 80 Ω, 80–85, 85–90, 90–95, 95–100, 100–105, 105–110, 110–115, 115–120 and above 120. The proce-

```
PROGRAM SORTANDCOUNT;

{* TO FIND FREQUENCY DISTRIBUTION AND PLOT
   BAR CHART OF A LIST OF VALUES *}

VAR  R,I,NOOFVALUES :INTEGER;
     N:ARRAY[0..10] OF INTEGER;

PROCEDURE INITIALISE;
BEGIN
FOR I:=0 TO 10 DO
N[I]:=0
END;

PROCEDURE READANDSORT;
BEGIN
WRITE('ENTER NO. OF R VALUES ');
READ(NOOFVALUES);
FOR I:=1 TO NOOFVALUES  DO
 BEGIN
  WRITE('R VALUE= ');
  READ(R);
  WRITELN;
  IF R<80  THEN  N[0]:=N[0]+1;
  IF (R>=80) AND (R<85) THEN  N[1]:=N[1]+1;
  IF (R>=85) AND (R<90) THEN  N[2]:=N[2]+1;
  IF (R>=90) AND (R<95) THEN  N[3]:=N[3]+1;
  IF (R>=95) AND (R<100)THEN  N[4]:=N[4]+1;
  IF (R>=100)AND (R<105)THEN  N[5]:=N[5]+1;
  IF (R>=105)AND (R<110)THEN  N[6]:=N[6]+1;
  IF (R>=110)AND (R<115)THEN  N[7]:=N[7]+1;
  IF (R>=115)AND (R<120)THEN  N[8]:=N[8]+1;
  IF (R>=120)AND (R<125)THEN  N[9]:=N[9]+1;
  IF R>125 THEN  N[10]:=N[10]+1
 END
END;

PROCEDURE PRINTABAR(P:INTEGER);
VAR X:INTEGER;
BEGIN
 FOR X:=1 TO P DO
  WRITE('*'); ;
WRITELN('  CLASS',I:3,' No.=',N[I]:4);
END;

BEGIN  {* MAIN PROGRAM *}
INITIALISE;
READANDSORT;
FOR I:= 0 TO 10 DO
 PRINTABAR(N[I])
END.
```

dure READANDSORT uses IF statements to sort the elements of array N to hold the counts in the various ranges. Procedure INITIALISE initialises all counts to zero before entering in values. Procedure PRINTABAR displays the component bars in the bar chart.

Here is an example of the program output obtained for a batch of a hundred resistors:

```
***************  CLASS  0 No.=  15
**  CLASS  1 No.=   2
****************  CLASS  2 No.=  16
*****  CLASS  3 No.=   5
******************  CLASS  4 No.=  19
*****************  CLASS  5 No.=  18
********  CLASS  6 No.=   8
****  CLASS  7 No.=   4
***  CLASS  8 No.=   3
*****  CLASS  9 No.=   5
*****  CLASS 10 No.=   5
```

Example 6.5: Sorting in numerical order

The following program may be used to sort numerical data in order. The numerical data to be sorted is first stored in array X using the procedure READINDATA. The procedure SORT, which itself includes the procedure SWOP, does the actual sorting by continually comparing the values held in two array elements and "swapping" them if one is less than (or equal to) the other. In this way data "bubbles" up or down until the entire array elements are sorted in order. The procedure DISPLAY displays the numbers in ascending order. If the display is required in descending order, either the greater than or the equal to symbol is used in the procedure SWOP, i.e

IF X[L]>=X[K] THEN

or change the FOR statement in the procedure DISPLAY to:

FOR I:=N DOWNTO 1 DO

Note that the maximum number of elements is determined (obviously) by the array size, 100 in our example.

```
PROGRAM DATASORT;

{Sorts numbers in ascending order}

VAR X:ARRAY[1..100] OF REAL;
    N:INTEGER;

PROCEDURE READINDATA;
VAR J:INTEGER;
BEGIN
WRITE('Enter values: ');
  FOR J:=1 TO N DO
    READ(X[J])
END;

PROCEDURE SORT;
VAR K,L:INTEGER;
 PROCEDURE SWOP;
  VAR TEMP:REAL;
  BEGIN
   IF X[L]<=X[K] THEN
    BEGIN
     TEMP:=X[L];X[L]:=X[K];X[K]:=TEMP
    END
  END;
BEGIN
 FOR K:=1 TO N-1 DO
  FOR L:=K+1 TO N DO
    SWOP
END;

PROCEDURE DISPLAY;
VAR I:INTEGER;
BEGIN
 FOR I:=1 TO N DO
  WRITE(X[I]:8:2)
END;

BEGIN { MAIN PROGRAM }
WRITE('Enter no. of values to be sorted: ');
READ(N);
 READINDATA;
 SORT;
 DISPLAY
END.
```

Example 6.6: Solution of equations, mesh and nodal analysis

This program uses a simple iterative method to solve linear simultaneous equations of the general type:

$$a_{11}x_1 + a_{12}x_2 + \cdots + a_{1N}x_N = c_1$$
$$a_{21}x_1 + a_{22}x_2 + \cdots + a_{2N}x_N = c_2$$
$$\vdots$$
$$a_{N1}x_1 + a_{N2}x_2 + \cdots + a_{NN}x_N = c_N$$

and is ideally suited, for example, for solving the mesh and nodal equations for resistive circuits.

Fig. 6.3 Flowchart illustrating basic steps in the iterative solution of a set of simultaneous equations

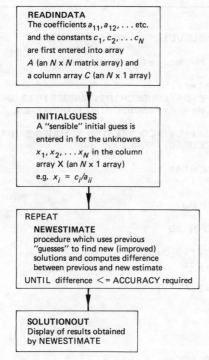

The form of the iterative process for obtaining a solution to a given accuracy is indicated in the flowchart diagram of fig. 6.3.

The basis of the method can be seen as follows where we consider, for example, the solution of three equations:

$$a_{11}x_1 + a_{12}x_2 + a_{13}x_3 = c_1 \tag{1}$$

$$a_{21}x_1 + a_{22}x_2 + a_{23}x_3 = c_2 \tag{2}$$

$$a_{31}x_1 + a_{32}x_2 + a_{33}x_3 = c_3 \tag{3}$$

$$\text{from (1)} \qquad x_1 = \frac{c_1 - a_{12}x_2 - a_{13}x_3}{a_{11}} \tag{4}$$

$$\text{from (2)} \qquad x_2 = \frac{c_2 - a_{21}x_1 - a_{23}x_3}{a_{22}} \tag{5}$$

$$\text{from (3)} \qquad x_3 = \frac{c_3 - a_{31}x_1 - a_{32}x_2}{a_{33}} \tag{6}$$

Thus a reasonable INITIALGUESS may be taken for the solutions as

$$x_1 = \frac{c_1}{a_{11}} \qquad x_2 = \frac{c_2}{a_{22}} \qquad x_3 = \frac{c_3}{a_{33}}$$

i.e. assuming $x_2 = x_3 = 0$ in (4), $x_1 = x_3 = 0$ in (5) and $x_1 = x_2 = 0$ in (6) respectively.

```
PROGRAM EQUATIONSOLVE;

CONST    N=3;   { N= No. of unknowns }

TYPE     SUBSCRIPT=1..N;
         AMATRIX=ARRAY[SUBSCRIPT,SUBSCRIPT] OF REAL;
         COLVECTOR=ARRAY[SUBSCRIPT] OF REAL;

VAR      A:AMATRIX; C,X:COLVECTOR; I:SUBSCRIPT;
         DXMAX,ACCURACY:REAL;

PROCEDURE READINDATA;
VAR I,J:SUBSCRIPT;
  BEGIN
   FOR I:=1 TO N DO
    BEGIN
     FOR J:=1 TO N DO
      BEGIN
       WRITE('A',I:2,J:2,':');
       READ(A[I,J])
      END;
    WRITE('C',I:2,':');
    READ(C[I]);
    WRITELN(' ')
    END
   END;

PROCEDURE INITIALGUESS;
VAR I:SUBSCRIPT;
BEGIN
 FOR I:=1 TO N DO
  X[I]:=C[I]/A[I,I]
END;

PROCEDURE NEWESTIMATE;
VAR I,K:SUBSCRIPT;
DX,NUMERATOR,NEWXVAL:REAL;
  BEGIN
   DXMAX:=0.0;
    FOR I:=1 TO N DO
     BEGIN
      NUMERATOR:=C[I];
       FOR K:=1 TO I-1 DO
        NUMERATOR:=NUMERATOR-A[I,K]*X[K];
       FOR K:=I+1 TO N DO
        NUMERATOR:=NUMERATOR-A[I,K]*X[K];
     NEWXVAL:=NUMERATOR/A[I,I];
     DX:=ABS(X[I]-NEWXVAL);
     IF DX>DXMAX THEN DXMAX:=DX;
     X[I]:=NEWXVAL
      END
   END;

PROCEDURE SOLUTIONOUT;
VAR I:SUBSCRIPT;
  BEGIN
   FOR I:=1 TO N DO
    WRITE('X',I:2,' = ',X[I]:8:4,'   ');
    WRITELN
  END;
```

```
BEGIN {*** MAIN PROGRAM ***}
WRITE('ENTER ACCURACY REQUIRED');
READ(ACCURACY);
WRITELN;
  READINDATA;
  INITIALGUESS;
    REPEAT
      NEWESTIMATE;
      SOLUTIONOUT
    UNTIL DXMAX<=ACCURACY
END.
```

These "guesses" are computed in the procedure INITIALGUESS and used to initialise the values for x_1, x_2, x_3 in the procedure NEWESTI-MATE.

The procedure NEWESTIMATE uses equations (4), (5), (6) to work out more accurate solutions. In the main program, the section NEWESTIMATE is used in a REPEAT loop to continually compute new estimates UNTIL the difference between two consecutive solutions differ by less than the required accuracy.

Obviously there are certain restrictions to obtaining a solution using this method, the most important being that there is a unique solution and that the method produces a convergent solution so that each progressive estimate becomes more accurate. In the program given above, solutions at each stage of iteration are given so you can see how many loops are required before the required accuracy (input by us in the main program) is achieved.

a) Application to Mesh Analysis

The program may be very easily applied to the mesh or nodal analysis of a circuit. Consider its application to the 3-mesh circuit of fig. 6.4.

The mesh equations for this circuit are:

$$(10+6+20) I_1 \qquad -6 I_2 \qquad 0 I_3 = 50$$
$$-6 I_1 + (32+12+15+6) I_2 \qquad +12 I_3 = 0$$
$$0 I_1 \qquad +12 I_2 + (40+12+60) I_3 = 125$$

so the values of the A matrix (coefficients of the currents I_1, I_2, I_3) are:

$$A_{11} = (10+6+20) = 36, \quad A_{12} = -6, \quad A_{13} = 0$$
$$A_{21} = -6, \quad A_{22} = (32+12+15+6) = 65, \quad A_{23} = 12$$
$$A_{31} = 0, \quad A_{32} = 12, \quad A_{33} = (40+12+60) = 112$$

Fig. 6.4 Circuit for mesh analysis problem

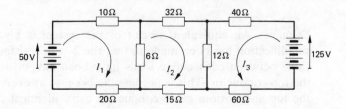

The values of the C matrix are:

$$C_1 = 50, \quad C_2 = 0, \quad C_3 = 125$$

All we need to do now is to run the program and enter in these values, first entering in the accuracy required for the solutions. Here is the display obtained:

```
ENTER ACCURACY REQUIRED 0.001

A 1 1:36
A 1 2:-6
A 1 3:0
C 1:50

A 2 1:-6
A 2 2:65
A 2 3:12
C 2:0

A 3 1:0
A 3 2:12
A 3 3:112
C 3:125

X 1 =    1.3889    X 2 =   -0.0778    X 3 =    1.1244
X 1 =    1.3759    X 2 =   -0.0806    X 3 =    1.1247
X 1 =    1.3755    X 2 =   -0.0807    X 3 =    1.1247
```

showing that we obtain the solution to within ±0·001 in 3 iterations:

$$I_1 = 1·376, \quad I_2 = -0·081, \quad I_3 = 1·125$$

b) Application to Nodal Analysis

Consider the following problem. Fig. 6.5a shows a diagram of a 2-wire d.c. distribution system fed at one end by a 240 V and at the other by a 250 V d.c. supply. The distributor cable has a resistance of 0·2 Ω/km per conductor and supplies power to the resistive loads connected between the conductors at the intervals shown in the diagram.

Write down the nodal equations of the system in the matrix form: $[G][V] = [I]$, where $[G]$ is a 6×6 conductance matrix and $[V]$ is a 6×1 column matrix containing the voltages developed across the 6 loads.

Use the program EQUATIONSOLVE to find the voltages V_1 to V_6. [Remember: set $N = 6$ for this case.]

Solution An equivalent circuit of the system is given by fig. 6.5b. Here simplification has been made in that the 2-wire resistances of the distributor cable between consecutive loads have been lumped together and placed in the top conductor. This is permissible because at every cross-sectional plane the top and bottom cable conductors carry identical current amplitudes.

Fig. 6.5 Circuits for nodal analysis problem

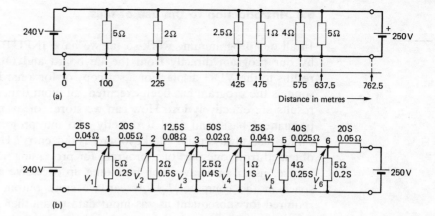

(a)

(b)

On applying the current law at the nodes of this equivalent circuit we have,

at node 1: $25(V_1 - 240) + 0 \cdot 2 V_1 + 20(V_1 - V_2) = 0$

i.e. $\qquad\qquad\qquad 45 \cdot 2 V_1 - 20 V_2 = 6000$

at node 2: $\qquad -20 V_1 + 33 V_2 - 12 \cdot 5 V_3 = 0$

at node 3: $\qquad -12 \cdot 5 V_2 + 62 \cdot 9 V_3 - 50 V_4 = 0$

at node 4: $\qquad\quad -50 V_3 + 76 V_4 - 25 V_5 = 0$

at node 5: $\qquad\quad -25 V_4 + 65 \cdot 25 V_5 - 40 V_6 = 0$

at node 6: $20(V_6 - 250) + 0 \cdot 2 V_6 + 40(V_6 - V_5) = 0$

i.e. $\qquad\qquad\qquad -40 V_5 + 60 \cdot 2 V_6 = 5000$

Thus in matrix form the equation for the system is

$$\begin{bmatrix} 45 \cdot 2 & \times 20 & 0 & 0 & 0 & 0 \\ -20 & 33 & -12 \cdot 5 & 0 & 0 & 0 \\ 0 & -12 \cdot 5 & 62 \cdot 9 & -50 & 0 & 0 \\ 0 & 0 & -50 & 76 & -25 & 0 \\ 0 & 0 & 0 & -25 & 65 \cdot 25 & -40 \\ 0 & 0 & 0 & 0 & -40 & 60 \cdot 2 \end{bmatrix} \begin{bmatrix} V_1 \\ V_2 \\ V_3 \\ V_4 \\ V_5 \\ V_6 \end{bmatrix} = \begin{bmatrix} 6000 \\ 0 \\ 0 \\ 0 \\ 0 \\ 5000 \end{bmatrix}$$

This equation was solved by feeding the above $[G]$ and $[I]$ data into the A array matrix, 6×6 values entered in, and the C matrix (6 values) respectively. The results obtained were:

$V_1 = 230 \cdot 918$ V $\qquad V_4 = 216 \cdot 613$ V

$V_2 = 221 \cdot 875$ V $\qquad V_5 = 225 \cdot 941$ V

$V_3 = 216 \cdot 281$ V $\qquad V_6 = 233 \cdot 184$ V.

6.4 Introduction to the use of files

In all our programming work so far we have INPUT data to be processed by our program directly from the keyboard and OUTPUT the required results to the VDU display for a softcopy or for a hardcopy to a printer. As soon as the program has been executed, all input data and any intermediate results are effectively lost. How can we store, for example, input data on a permanent basis and feed it directly into our program(s) when required, rather than always using the keyboard as our entry? How can we store data obtained from one part of a program for processing in another (without, of course, using an array which may take up excessive and valuable memory space in the computer)? How can we store output data which may be required for subsequent use as input data to another program?

These problems may be overcome by storing input, output and intermediate "transferable" data in files in an auxiliary memory device, such as a floppy disc or magnetic tape, coupled into the computer system. We can then, for example, prepare input data independently, save it on disc, and read in the data when required in program execution. Likewise to save intermediate data or output data we can direct it to be written onto disc.

PASCAL files consist of data of the same type (e.g. real, integer, char) stored in sequence in auxiliary memory. PASCAL handles only these forms of sequential files, where individual items of data can only be accessed in the order in which they are stored. A Pascal file can only be examined one element at a time, starting from the beginning. Unlike an array, a file cannot be accessed randomly. File size is not specified and a file may be allowed to expand (subject to available memory) to any size.

6.5 File declaration

All files used in our programs must be declared. The syntax for file declaration is

TYPE   file-type=FILE OF data-component-type;

VAR   file-identifier: file-type;

which may be combined into the short-hand form:

VAR   file-identifier: FILE OF data-component type;

For example,

TYPE   WHOLENOFILE = FILE OF INTEGER;

DECIMALNOS = FILE OF REAL;

LETTERS = FILE OF CHAR;

VAR   AGES,POPULATION,REFNOS: WHOLENOFILE;

TEMPS,POWERS,RVALUES: DECIMALNOS;

NAMES,ADDRESSES: LETTERS;

or in the short-hand version:

VAR   AGES,POPULATION,REFNOS: FILE OF INTEGER;

TEMPS,POWERS,RVALUES: FILE OF REAL;

NAMES,ADDRESSES: FILE OF CHAR;

In addition, provision must also be made to set up a channel of communication for every external file declared in the program. An external file is one which will exist either before and/or after program execution, e.g. files on disc which are used to store input and output data. This channel of communication between computer and file obviously depends on the system being used.

In the majority of PASCAL mainframe computer implementations, external file identifiers are included alongside the standard PASCAL INPUT and OUTPUT files identifiers in the program heading. For example,

PROGRAM   TESTANALYSIS(INPUT,OUTPUT,INDATA,OUTRESULTS);

would be the form a program heading might take when the program is to take input data from the keyboard and an external file and output to both a VDU (or printer) and an external file:

INPUT is the standard PASCAL file used to indicate that data external to the program is to be READ from the keyboard

OUTPUT is the standard PASCAL file used to indicate that data is to be WRITtEn to the screen (VDU) or a printer

INDATA is an identifier for an external file, e.g. containing input data stored in an auxiliary memory device

OUTRESULTS is an identifier for an external file, e.g. to be set up by the program to store output data in an auxiliary memory device.

In PASCAL implementations where program parameters are not included in the program heading (as in this book), communication between computer and external file source/destination is set up, including additional information, in the RESET and REWRITE statements used in PASCAL to, respectively, make a source file ready for READing from and creating or opening a file ready for WRITEing to.

6.6 Reading from and writing to files: RESET and REWRITE statements

In this section we consider how input data may be READ from an external file and how we can create and WRITE data output to an external file. To aid the explanation we assume that we have declared the file FILETEST and variables F, F1, F2, F3, $\cdots$ of the same data type as the file data items.

For example,

> VAR   FILETEST: FILE OF REAL;
> F,F1,F2,F3 :REAL;

a) To READ data from a file

We must always precede the read operation by the statement;

> RESET(file-identifier);

This call of the standard PASCAL procedure RESET positions to the first item of data in the file ready for reading.

The first item can then be read from the file using the statement:

> READ(file-identifier,F);

i.e. the variable F of the same data type as the file data is assigned the value of the first data item in the file.

A second

> READ(file-identifier,F);

statement assigns F the second data item of the file to F, and so on.

We can, of course, read more than one file item using a single READ statement. For example,

> READ(FILETEST,F1,F2,F3);

would assign F1, F2, F3 the next three consecutive items of the file.

For systems not including program parameters in the program heading, a channel of communication between computer and file source is set up by including additional information in the RESET statement. The exact form this takes is obviously implementation-dependent but typically takes a form similar to the following:

> RESET(file-identifier,'B:file-name-on-disc');

where the location of the file and its name are included within quotes. In the above example the file to be read is on disc in disc drive B and indicated by B: The name of the file must also be quoted.

b) Pictorial representation of RESET, READ and the End of File, EOF, function

A pictorial illustration of the action of the RESET and READ statements is shown in fig. 6.6. The RESET statement creates a window into the file allowing the first item of file data to be "seen", i.e. loaded from the external memory device on which the file is stored to the file buffer, which is the memory interfacing directly with the computer and temporarily holding data before (and after) processing.

Fig. 6.6 Pictorial representation of sequential file and RESET, READ statements for reading from files

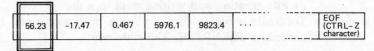

(a) RESET(FILETEST);   sets "window" to first item in file
     READ(FILETEST,F); assigns F value of first item, i.e. F:=56.23

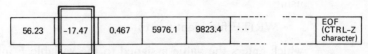

(b) Once a file item is READ, the "window" moves to the next item.
     Thus a second READ statement:
       READ(FILETEST,F);   assigns F value of second file item and if followed by
       READ(FILETEST,F1,F2,F3);
     assigns F1, F2, F3 the next three data items respectively, i.e.
     F1:=0.467 , F2:=5976.1 , F3:=9823.4

All PASCAL files should be terminated by the end-of-file EOF marker (in practice, the EOF marker is a CTRL-Z character). We can then utilize the EOF function. EOF (file-identifier) is a Boolean function which returns TRUE if the file window or pointer is positioned at the end of the file, i.e. on the EOF character, but otherwise returns the value FALSE.

The use of the EOF function, RESET and READ is illustrated in the following example.

Example 6.7 This program reads in numerical data from a file stored on disc, works out the mean, root mean square and standard deviation of the file data, and displays the results.

```
PROGRAM STATISTICS2;

VAR   DATA:TEXT;
      C,SUM,SUMSQ:REAL;
      N:INTEGER;

BEGIN
  RESET(DATA,'B:CAPACITR.TXT');
  SUM:=0;SUMSQ:=0;N:=0;
  WHILE NOT EOF(DATA) DO
    BEGIN
      READ(DATA,C);
      WRITE(C:8:2);
      SUM:=SUM+C;
      SUMSQ:=SUMSQ+C*C;
      N:=N+1
    END;
WRITELN('MEAN = ',SUM/N:8:2);
WRITELN('RMS VALUE = ',SQRT(SUMSQ/N):8:2);
WRITE('STANDARD DEVIATION = ');
WRITELN(SQRT( SUMSQ/N-SUM*SUM/(N*N)):8:4)
END.
```

c) **File creation and writing data to a file**

To create a file ready for writing in data, we call the standard PASCAL procedure REWRITE. The statement

REWRITE(file-identifier);

initializes a file ready for writing to. Data may then be written to the file using

WRITE(file-identifier,F);

which enters the value assigned to the variable F into the file.

WRITE(file-identifier,F1,F2,F3);

would effect the entry in sequence of the three data values assigned to F1, F2, F3, respectively.

For implementations of PASCAL not using program parameters in the program heading, additional information must be included in the RE-WRITE statement so as to open a channel of communication for writing from the computer to the external memory device. For example,

REWRITE(FILETEXT,'B:FILETEST.002');

would initialize a file named FILETEST.002 on a disc located in disc-drive B.

Example 6.8 This program illustrates:
(i) the mechanism of creating and writing to a file (see procedure WRITEFILE)
(ii) reading data from the file and storing each item in an array (see procedure READFILE)
(iii) Checking the above two operations (see procedure CHECK).

```
PROGRAM FILETEST1;

VAR   NUMBERS1:TEXT;
      I        :INTEGER;
      A: ARRAY[1..20] OF INTEGER;

PROCEDURE WRITEFILE;
{To open a file and write data to
 this file. In this example the file is
 to be stored in disc in drve B and
 the file name is NUMBERS1.001 }

BEGIN
 REWRITE(NUMBERS1,'B:NUMBERS1.001');
  WRITELN(NUMBERS1,20);
    FOR I:=1 TO 20 DO
     WRITE(NUMBERS1,I)
END;
```

```
PROCEDURE READFILE;
{To read in the data stored in file
 NUMBERS1.001 on the disc in drive B}
VAR N:INTEGER;
BEGIN
 RESET(NUMBERS1,'B:NUMBERS1.001');
 READ(NUMBERS1,N);
 FOR I:=1 TO N DO
   READ(NUMBERS1, A[I] )
END;

PROCEDURE CHECK;
BEGIN
 FOR I:=1 TO 20 DO
   WRITE(A[I]:6)
END;

BEGIN { Main program }
 WRITEFILE;
 READFILE;
 CHECK
END.
```

On running the program you will obtain a display of the first twenty integer numbers, i.e. 1 2 3 · · · 20.

> WRITEFILE writes to the external file, the number 20 and then the first twenty numbers.
>
> READFILE reads this file and stores the file data in array A.
>
> CHECK outputs the array element values to the screen.

6.7 TEXT files

When data is written into FILES OF REAL and INTEGER, the individual items are normally transferred and stored in "internal" binary code form that can be immediately interpreted and used by the computer without any need of translation. Files where the data is stored in internal code are known as **binary files**. Data in a binary file cannot be used directly to display its meaning to the outside world. Binary file data must first be translated into a more "universal" code, normally the ASCII code, which is then used to effect the display on the screen of the VDU, or to drive a printer, or any other peripheral device connected to the system.

Thus, in addition to binary files, PASCAL also utilizes files—known as **text files**—to provide a direct means of communication between the computer and its peripherals. A text file is essentially a file that stores codes representing a sequence of characters. Text files, however, have an additional property not shared by binary files. They are sequential files of character sequences considered as divided up into lines. Each line is terminated by a *line-separator character*. This character can only be generated in the file creation operation by using the standard PASCAL procedure WRITELN. For example,

WRITELN(file-identifier);

generates a line-separator character.

The line-separator cannot be read in the normal way but is detected using the standard PASCAL function. The end-of-line (EOLN) function:

EOLN(file-identifier)

is a Boolean function which returns TRUE if and only if the current character being read from a text file is the line-separator character. For all other characters including space, the value returned by EOLN(identifier) is FALSE. Thus, for example, to read and display the first line of a text file, FILETEXT, we could use:

```
RESET(FILETEXT);
WHILE NOT EOLN(FILETEXT) DO
BEGIN
    READ(FILETEXT,F);
    WRITE(F)
END;
```

Text files are declared in the VAR section as follows:

VAR   file-identifier: TEXT;

Text files are opened for writing to in the normal way using

REWRITE(file-identifier);

or by   REWRITE(file-identifier,'D:file-name');

the latter being used for implementations not declaring formal parameters in the program heading.

Text files are opened for reading, again in the normal way, using

RESET(file-identifier);

or by   RESET(file-identifier,'D:file-name');

Data may now be read from the text file and automatically converted in the READ statement into the required data type. For example:

a) Data of type INTEGER

Suppose we have declared a text file and variable of type INTEGER in the VAR section, e.g.

VAR   NUMBERFILE: TEXT;
           N,N1,N2,N3 :INTEGER;

then the statement,

READ(NUMBERFILE,N);

assigns to INTEGER variable N the INTEGER value corresponding to characters currently being read from the text file. The computer automatically works out what integer values are represented by the character sequence being read. All preceding blanks (spaces) and end-of-line markers are skipped. If the integer read has an absolute value greater than MAXINT then the run-time error "Number too large" is displayed and execution stopped. If the first character read after spaces and EOLN characters have been skipped is not a digit or a + or − sign, then the run-time error "Number expected" will be displayed and execution again stopped.

READ(NUMBERFILE,N1,N2,N3);

assigns respectively to the three INTEGER variables N1, N2, N3 the INTEGER values corresponding to three consecutive character sequences of the text file, with any interleaving spaces or EOLN characters being skipped.

b) Data of type REAL

Suppose we have made the declaration

VAR   DECIMALDATA: TEXT;
        R,R1,R2,R3 :REAL;

then the statement

READ(DECIMALDATA,R);

assigns the REAL variable R the REAL value corresponding to the series of characters currently being read from the text file. All leading spaces and EOLN characters are skipped and, as for INTEGERS, the first character afterwards must represent either a digit or a + or − sign.

Likewise,

READ(DECIMALDATA,R1,R2,R3);

reads three consecutive items from the textfile and converts them to REAL values which are assigned to the variables R1, R2 and R3.

c) Data of type CHAR

Suppose we have made the declaration

VAR   CHARACTERFILE: TEXT;
        CHA: CHAR;

then the statement

READ(CHARACTERFILE,CHA);

reads the current character from the file and this is assigned to the variable CHA. If the window of the file is positioned on an end-of-line character, then the EOLN function EOLN(CHARACTERFILE) returns the value TRUE. Thus, when a READ operation is next performed on the file, the file window will be positioned at the start of a new line.

6.8 The applications of files in programs

The following examples illustrate some of the important applications of files in PASCAL programs: for example, for providing permanent records on file of input and output data and data such as tables of special functions, etc. that can be referenced when required by a program.

Remember:

1 Files must be declared in the VAR (or TYPE and VAR) sections of the program, e.g.

```
VAR     GENERALDATA:   TEXT;
        RVALUES,FUNCTIONTABLE:   FILE OF REAL;
        UNITNUMBERS:   FILE OF INTEGER;
```

2 Before reading from a file, the file must be opened using

RESET(file-identifier);

with, for certain PASCAL implementations, additional information as to the location of the file and the name under which it is stored.

The file can then be read sequentially starting with the first data item of the file using

READ(file-identifier,F);

where F is a variable of the same data type as the file items and which is also declared in the VAR section Note that files, unlike arrays, can only be read in order starting from the first item; they cannot be accessed randomly.

3 To create a file for writing to, use

REWRITE(file-identifier);

plus, where the implementation requires, information as to the file location and the name under which the data items are to be stored.

Data can then be entered into the file using

WRITE(file-identifier,F);

Example 6.9:   To process data on file and output results to a new file
This program reads numerical data from an existing file, FILE1, which is stored on a disc in disc drive B under the name DATA1111 . TXT. It creates a new file, FILE2 (stored as DATA2222.TXT on disc drive B), and writes to this file all items of data from FILE1 which lie in the range 80 to 110.

Try running the program, having first prepared FILE1 as an external file, either as TEXT file using for example your text editor on a FILE OF REAL or INTEGER where in this case you must prepare FILE1 using a program to write your file.

```
PROGRAM DATARANGESELECT;

VAR FILE1,FILE2:TEXT;
    N:REAL;

BEGIN
 RESET(FILE1,'B:DATA1111.TXT');
 REWRITE(FILE2,'B:DATA2222.TXT');
  WHILE NOT EOF(FILE1)  DO
    BEGIN
     READ(FILE1,N);
       IF (N>=90) AND (N<=110) THEN
          WRITE(FILE2,N)
    END;
{ Display of file, FILE1: }
 RESET(FILE1,'B:DATA1111.TXT');
 RESET(FILE2,'B:DATA2222.TXT');
 WRITELN('FILE 1:');
 WRITELN('-------');
 WHILE NOT EOF(FILE1) DO
  BEGIN
    READ(FILE1,N);  WRITE(N:6:0)
  END;
 WRITELN;
{ Display of file, FILE2 }
 WRITELN('FILE 2:');
 WRITELN('-------');
 WHILE NOT EOF(FILE2) DO
  BEGIN
    READ(FILE2,N);  WRITE(N:6:0)
  END
END.
```

Example 6.10: Sideband components of an FM wave

This program displays the frequency components and amplitudes of a sinusoidally modulated FM (frequency modulated) wave of the form

$$v = \cos\left(\omega_c t + \frac{f_d}{f_m}\sin\,\omega_m t\right)$$

where $f_c = \omega_c/2\pi$ is the carrier frequency

$f_m = \dfrac{\omega_m}{2\pi}$ is the signal frequency

f_d = frequency deviation.

The amplitude of a sideband component at the frequency $f_c \pm nf_m$, $n = 0, 1, 2, 3, \ldots$ are given by the value of the Bessel function $J_n(f_d/f_m)$; see table of values below. This table is stored in an external file and loaded into the program for use by the procedure READBESSEL.

Order	Values of Bessel function of first kind $J_n(x)$ for					
n	$x = 1$	2	3	4	5	6
0	0·7652	0·2239	−0·2601	−0·3971	−0·1776	0·1506
1	0·4401	0·5767	0·3391	−0·0660	−0·3276	−0·2767
2	0·1149	0·3528	0·4861	0·3641	0·0466	−0·2429
3	0·0196	0·1289	0·3091	0·4302	0·3648	0·1148
4	0·0025	0·0340	0·1320	0·2811	0·3912	0·3576
5	0·0003	0·0070	0·0430	0·1321	0·2611	0·3621
6	0·0002	0·0012	0·114	0·0491	0·1310	0·2458

```
PROGRAM P25;

VAR BESSELFN:TEXT;
    N,X      :INTEGER;
    J:ARRAY[0..6,1..6] OF REAL;
    FC,FD,FM,M    :REAL;


PROCEDURE READBESSEL;
BEGIN
RESET(BESSELFN,'B:BESSELFN.TXT');
FOR N:=0 TO 6 DO
   FOR X:=1 TO 6 DO
      READ(BESSELFN,J[N,X])
END;

BEGIN
WRITE('ENTER CARRIER,MODULATING AND');
WRITELN(' DEVIATION FREQUENCIES');
READ(FC,FM,FD);
READBESSEL;
X:=ROUND(FD/FM);
WRITELN('****************************************');
WRITE('COMPONENTS IN FREQUENCY SPECTRUM');
WRITELN('    OF F.M. WAVE ');
WRITELN('****************************************');
WRITELN('FREQUENCY          AMPLITUDE');
WRITELN('------------------------------------');
WRITELN('FC=   ',FC:9,'          ',J[0,X]);
WRITELN('FC+/-',FM:9,'          ',J[1,X]);
WRITELN('FC+/-',2*FM:9,'        ',J[2,X]);
WRITELN('FC+/-',3*FM:9,'        ',J[3,X]);
WRITELN('FC+/-',4*FM:9,'        ',J[4,X]);
WRITELN('FC+/-',5*FM:9,'        ',J[5,X]);
WRITELN('FC+/-',6*FM:9,'        ',J[6,X]);
WRITELN('------------------------------------')
END.
```

Here is an example of the results output on running the program for the case $f_c = 90$ MHz, $f_m = 15$ kHz, $f_d = 75$ kHz.

```
P25B:P25
ENTER CARRIER,MODULATING AND DEVIATION FREQUENCIES
90E6 15E3 75E3
****************************************
COMPONENTS IN FREQUENCY SPECTRUM      OF F.M. WAVE
****************************************
FREQUENCY              AMPLITUDE
----------------------------------------
FC=     9.00E+07          -1.77600E-01
FC+/- 1.50E+04           -3.26700E-01
FC+/- 3.00E+04            4.66000E-02
FC+/- 4.50E+04            3.64800E-01
FC+/- 6.00E+04            3.91200E-01
FC+/- 7.50E+04            2.61100E-01
FC+/- 9.00E+04            1.31000E-01
----------------------------------------
```

Fig. 6.7 Best-fit straight line

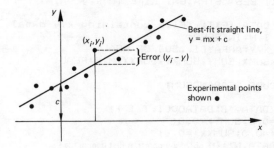

Example 6.11: "Best-fit" straight line

This program may be used to find the best-fit straight line for a number of experimental points (x, y) logged into an external file. The best-fit straight line (see fig. 6.7):

$$y = mx + c$$

where m = gradient of line
c = intercept on y axis

is found by a least-squares method which is designed to minimise the sum of the errors between each experimental point and the best-fit straight line point, i.e. for n (x, y) points,

$$\sum_{i=1}^{n} (y - y_i)^2 \quad \text{is minimised}$$

Applying this criterion gives:

$$m = \frac{n\sum x_i y_i - \sum x_i \sum y_i}{n\sum x_i^2 - \left(\sum x_i\right)^2}$$

$$c = \frac{1}{n}\left[\sum y_i - m\sum x_i\right]$$

The procedure READANDSUM accomplishes the $\sum x_i y_i$, etc. calculations, first drawing the $x-y$ data from file and loading x, y values into arrays X and Y respectively for storage. Actually there is no need to use arrays. Try modifying the procedure to eliminate them—this will save valuable storage space and may even be essential if the $x-y$ data file is very large.

The procedure BESTLINE computes and displays the gradient m and intercept c of the best-fit straight line.

```
PROGRAM BESTLINE;

{TO FIT BEST STRAIGHT LINE TO x,y DATA}

VAR     DATA:TEXT; {File containing x,y data}
        I,N:INTEGER;
        X,Y:ARRAY[1..50] OF REAL;
        SUMX,SUMY,SUMXY,SUMXX,M,C:REAL;

PROCEDURE READANDSUM;
BEGIN
RESET(DATA,'B:DATA0001.TXT');
SUMX:=0.0;SUMY:=0.0;
SUMXY:=0.0;SUMXX:=0.0;
READ(DATA,N); {N, no. of x-y points, is first item in file}
WRITELN(N:4,'...NO. OF EXPT. POINTS');
WRITELN;
WRITELN('    x          y   ');
WRITELN('-------------------------');
FOR I:=1 TO N DO
 BEGIN
  READ(DATA,X[I],Y[I]);
  WRITELN(X[I]:6:2,Y[I]:12:2);
   SUMX:=SUMX+X[I];
   SUMY:=SUMY+Y[I];
   SUMXY:=SUMXY+X[I]*Y[I];
   SUMXX:=SUMXX+X[I]*X[I]
 END;
WRITELN('-------------------------')
END;

PROCEDURE BESTLINE;
BEGIN
M:=(N*SUMXY-SUMX*SUMY)/(N*SUMXX-SUMX*SUMX);
C:=(SUMY-M*SUMX)/N;
WRITELN('VALUES OF m AND c FOR BEST-FIT STRAIGHT LINE');
WRITELN(' ARE   m = ',M:6:3,' c = ',C:6:3)
END;

BEGIN
READANDSUM;
BESTLINE
END.
```

Here is a test run of the program for 20 $x-y$ values stored in the test file DATA, the actual file being on disc and named there as DATA0001.TXT.

```
20...NO. OF EXPT. POINTS

        x               y
    ----------------------------
        0.00            1.30
        1.00            1.50
        2.00            2.30
        3.50            2.90
        5.00            4.00
        6.70            5.00
        8.10            6.20
        9.30            8.30
       10.00           10.50
       12.00           13.30
       13.30           15.60
       16.60            9.40
       20.10           11.80
       24.50           27.50
       27.70           30.60
       30.90           34.00
       34.00           37.90
       37.30           42.90
       41.40           47.50
       44.20           51.80
    ----------------------------
VALUES OF m AND c FOR BEST-FIT STRAIGHT LINE
ARE   m =  1.160  c = -1.938
```

Exercises 6

6.1 Construct a program which consists of the following two procedures:
ENTERDATA   which reads from the keyboard the numbers and unit cost of up to
20 items and stores the information in an array.
DISPLAYTABLE   which displays a table consisting of item number, number of
items, unit cost, sub-total (number × unit cost) and the complete total cost of all
items.

6.2 Construct a program which accomplishes the following: reads 32 rows of 4 values per
row from the keyboard and stores the values in an array; displays each row and
average value of each row; finally displays the average value of each column.

6.3 Construct a program which will allow up to 100 entries of real numbers to be entered
from the keyboard and on selection displays one of the following:
   the AVERAGE value
   the RMS (root mean square) value
   the MAXIMUM and MINIMUM value
   the numbers sorted in ascending order.

Fig. 6.8 Twin-tee
network

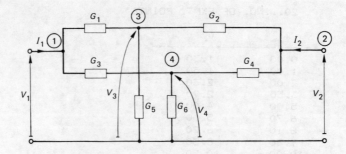

6.4 The nodal equations for the twin T-network of fig. 6.8 are

$$(G_1 + G_3)V_1 + \quad 0 \quad V_2 \quad\quad -G_1\,V_3 \quad\quad -G_3\,V_4 = 0$$
$$0 \quad V_1 + (G_2 + G_4)V_2 \quad -G_2\,V_3 \quad\quad -G_4\,V_4 = 0$$
$$-G_1\,V_1 \quad\quad -G_2\,V_2 + (G_1 + G_2 + G_5)V_3 \quad\quad 0 \quad V_4 = 0$$
$$-G_3\,V_1 \quad\quad -G_4\,V_2 \quad\quad 0 \quad V_3 + (G_3 + G_4 + G_6)V_4 = 0$$

where V_1, V_2, V_3, V_4 are the node voltages
$G_1, G_2, G_3, G_4, G_5, G_6$ are the conductances ($G = 1/R$)
I_1, I_2 are the input currents.

Write a program to solve these equations using the following procedures:
ENTERDATA to feed in conductance data and input current data into a conductance and current arrays.
ACCURACY to feed in accuracy to which node voltages are required.
SOLVE to solve the equations to within a given accuracy.
DISPLAYRESULTS to display results and number of iterations which are used before stated accuracy is achieved.

6.5 Construct a program that
a) writes the following table of real values to an external file:

9·1	54·6	27·8	42·7	15·9	16·23	67·8	8·88
14·2	0·72	49·6	0·4	69·2	37·84	14·6	11·2
62·3	50	43·2	98·1	15·6	32·43	86·3	57·7

b) reads the file and displays all the values on the screen and also determines the average and maximum and minimum values.

6.6 Construct a program that reads INTEGER data from an external file which contains numbers in the range 0 to 100 and counts the number of values in the following bands: 0 to 20, 21 to 30, 31 to 40, 41 to 49, 50 to 59, 60 to 69, above 70.

7 Some Further Topics: Packed Arrays, Records and Sets

7.1 Introduction and summary

In this final chapter some further programming concepts available in PASCAL—packed arrays and strings, records and sets—are introduced.

Packed arrays save valuable memory space and are particularly useful for storing and handling strings (e.g. character data such as names, words, symbols, etc.).

Records as the name suggests allow us to define a structure for records which may contain data of different types. This is particularly useful for general programming work where records of mixed data types are invariably required. Finally, sets and how they are applied in PASCAL are considered.

7.2 Packed arrays and strings

In order to economise on valuable computer memory space, especially important when handling large arrays of CHAR data, it is possible to pack several items of data into single storage locations.

Packing has little benefit for REAL and INTEGER data types since these are already stored efficiently. However, packing for CHAR, BOOLEAN, enumerated and sub-range data types does save memory space, typically by a factor of four for CHAR data type. Packing makes no difference to the meaning of a program nor does it affect its results. It is still possible to access individual elements in a packed array but this is a more time-consuming process. Thus in general, packing makes economies in memory space at the expense of slower access times.

In PASCAL, packing is achieved simply by prefixing **ARRAY**, or RECORD for packed RECORDS, by the reserved word **PACKED**. Thus a packed array may be defined as follows:

TYPE   *array-identifier* = PACKED ARRAY[1 . . 30] OF data type;

For example,

TYPE   NAME = PACKED ARRAY[1 . . 20] OF CHAR;
            ADDRESS = PACKED ARRAY[1 . . 50] OF CHAR;
VAR   NAMELIST: ARRAY[1 . . 100] OF NAME;
           ADDRESSLIST: ARRAY[1 . . 100] OF ADDRESS;

NAME and ADDRESS are defined as PACKED ARRAYS in the TYPE program section, whilst in the VAR section NAMELIST and ADDRESS-LIST are defined respectively as arrays of 100 elements of type NAME and ADDRESS. Thus, for example, each array element NAMELIST[I] is itself a "packed" array of 20 CHAR elements.

Packing is particularly important for the economic storage of character strings. The term **string** in PASCAL is used to denote a packed array of characters or, as we have already seen, the characters within the single quotes ' ' used in WRITE statements. Packed arrays of CHAR, i.e. strings, have a number of useful properties:

1. String variables can be assigned values in a single assignment statement, e.g. assuming the VAR declaration

VAR   SURNAME: PACKED ARRAY[1 .. 15] OF CHAR;

then   SURNAME:= 'MEADOWS R G   ';

assigns the value 'MEADOWS R G   ' to the string variable SURNAME. *Note*: the string is defined as 15 characters long and so, in all assignments to SURNAME, 15 characters must be included. If the actual number is less, fill in up to 15 using blanks, i.e. spaces, as indicated above.

2. In many implementations of PASCAL, string variable value can be output at one go, e.g.

WRITELN(SURNAME);

displays the value assigned to SURNAME.
Note: although a string may be output at one go, a string variable can normally only be READ one character at a time, e.g.

FOR I:= 1 TO 15 DO
    READ NAME[I];

Some PASCAL implementations, however, do allow a string variable to be input (READ) as a complete entity.

3. Strings of equal length can be compared using the six comparison operators. This is particularly useful in sorting lists of names, etc. in lexicographic (alphabetic) order (see example 7.2).

Example 7.1 This simple program illustrates how a string may be input (READ) from the keyboard and how its value can be subsequently displayed. Note that the string variable SURNAME (a packed array of 15 CHAR elements) is READ in character by character and that blanks (spaces) are automatically inserted to make up the total string length to 15.

To the user, however, this is not apparent. Just type in the name not exceeding 15 characters and terminate with a Carriage return or Enter key press. The latter will be interpreted as an EOLN, end-of-line, character.

```
PROGRAM PACKEDARRAYEX1;

TYPE  NAME =PACKED ARRAY[1..15] OF CHAR;

VAR
     LENGTH,I : INTEGER;
     SURNAME  : NAME;
BEGIN
 WRITELN('ENTER SURNAME');
 (* characters READ one at a time *)
 I:=1; LENGTH:=0; (* Initialization *)
 READLN; (* READ set to beginning of line *)
 WHILE NOT EOLN  DO
  BEGIN
   READ(SURNAME[I]);
   I:=I+1;
   LENGTH:=LENGTH+1
  END;
 IF LENGTH<15 THEN (* Filling in using spaces *)
   FOR I:=LENGTH+1 TO 15 DO
     SURNAME[I]:=' ';
 (* Displaying SURNAME letter by letter *)
 FOR I:=1 TO 15 DO
   WRITE(SURNAME[I]);
 WRITELN;
 (* Displaying SURNAME in one go *)
 WRITELN(SURNAME)
END.
```

The EOLN function is a standard Boolean function which will examine the next input character and return the value FALSE if the character is other than an EOLN character. If an EOLN character is present, the function returns a TRUE value.

Thus as soon as the EOLN character is met, the WHILE loop will be exitted and then, if need be, spaces are inserted to bring up the total to 15.

Example 7.2 This program utilizes the property that strings of fixed length may be compared—remember that CHAR data has its associated number code, see section 3.7—and therefore sorted in alphabetic order.

The array COMPNAMES is declared as an array of 10 elements, each of which is a packed array or string of 20 characters.

The procedure READNAMES allows us to enter in from the keyboard X names (maximum in this case 10, although obviously this can be increased by increasing the upper dimension limit of COMPNAMES).

The procedure SORTNAMES utilizes the fact that we can compare strings of fixed length and a bubble-sort type method to sort the names in alphabetic order.

The procedure DISPLAYNAMES displays the complete list of names, i.e. the elements of the array COMPNAMES. In the program it is used to display the list before and after sorting.

```
PROGRAM ALPHABETICSORT;

TYPE  NAME =PACKED ARRAY[1..20] OF CHAR;

VAR   COMPNAMES : ARRAY[1..10] OF NAME;
      X,N        :INTEGER;

PROCEDURE READNAMES(X:INTEGER);
VAR   COMP:NAME;
      LENGTH,I,N :INTEGER;
BEGIN
 FOR N:=1 TO X DO
  BEGIN
   WRITELN('ENTER COMPONENT NAME');
   I:=1; LENGTH:=0;
   READLN;
   WHILE NOT EOLN  DO
    BEGIN
     READ(COMP[I]);
     I:=I+1;
     LENGTH:=LENGTH+1;
     END;
   IF LENGTH<20 THEN
    FOR I:=LENGTH+1 TO 20 DO
     COMP[I]:=' ' ;
     COMPNAMES[N]:=COMP
  END
END;

PROCEDURE DISPLAYNAMES(X:INTEGER);
VAR I:INTEGER;
BEGIN
 FOR I:=1 TO X DO
  WRITELN(COMPNAMES[I]);
WRITELN
END;

PROCEDURE SORTNAMES(X:INTEGER);
VAR  K,L  :INTEGER;
     TEMP :NAME;
BEGIN
  FOR K:=1 TO X-1 DO
   FOR L:=K+1 TO X DO
    BEGIN
     IF COMPNAMES[L] < COMPNAMES[K] THEN
      BEGIN
       TEMP:=COMPNAMES[L];
       COMPNAMES[L]:=COMPNAMES[K];
       COMPNAMES[K]:=TEMP
      END
    END
END;

BEGIN {*** Main Program ***}
WRITELN('ENTER NO. OF NAMES TO BE READ');
READ(X);
READNAMES(X);
WRITELN('LIST OF NAMES IS:');
DISPLAYNAMES(X);WRITELN;
SORTNAMES(X);
WRITELN('LIST IN ORDER IS:');
DISPLAYNAMES(X)
END.
```

7.3 Records

In many applications it is very useful to be able to organise a record consisting of a fixed number of items which can be of different data types. For this purpose, PASCAL provides the **RECORD** data structure which consists of a fixed number of components, known as **fields**. Unlike arrays these fields do not have to be of the same data type.

The form of a RECORD definition is

TYPE   *record-identifier* = RECORD
                          field-identifier:*data type*;
                      . . .   .
                          field-identifier:*data type*;
                          END;

where the field components are contained between the reserved words RECORD and END. For example,

```
TYPE    COMPONENT = RECORD
                    NAME:PACKED ARRAY[1 . . 25] OF CHAR;
                    VALUE:REAL;
                    RATING:1 . . 100
                    END;
VAR     R,L,C,DIODE:  COMPONENT;
```

i.e. COMPONENT is defined as a RECORD with 3 fields: NAME, VALUE and RATING. Four variables R, L, C and DIODE of type COMPONENT are also declared in the VAR section.

Individual field components of a record can be accessed using the record identifier followed by a full stop . and then the field identifier. For example, the fields of the record variable R could be assigned as follows:

```
R.NAME:= 'Wire-wound resistor      ';
R.VALUE:= 100.0   ;
R.RATING:= 10;
```

For added convenience, PASCAL provides the **WITH** statement which enables individual record fields to be referenced without using the record variable identifier, provided that this has been specified in the first line of the WITH statement. The WITH statement takes the form:

```
WITH record-variable-identifier DO
  BEGIN
    field-identifier:= · · ·
  END;
```

For example,

```
WITH C DO
  BEGIN
  NAME = 'Tantalum electrolytic      ';
  VALUE:= 1000{*micro-farads*};
  RATING:= 20{*volts dc*}
END;
```

which assigns values to the fields of the record C.

```
WITH C DO
  WRITELN(NAME,VALUE,RATING);
```

may be used to display the fields assigned to C.

```
WITH C DO
  BEGIN
  READLN;
  FOR I:= 1 TO 25 DO
    READ NAME[I];
  READLN;READLN(VALUE);
  READLN(RATING)
END;
```

allows us to enter the field data from the keyboard.

In addition to containing a fixed number of fields, records may also have a **variant** part. This allows greater flexibility in that the number and types of certain fields can be varied to suit particular requirements. For example, the RECORD COMPONENT we used to introduce records consisted of three fixed fields—NAME, VALUE, RATING. Suppose we wish to add information giving supplier details such as address, price, delivery if available in the UK, or if not, just the countries where the component is available. Using the concept of variant records, we could proceed as follows:

```
TYPE     AVAILABILITY = (UK,NOTUK);
         COMPONENT = RECORD
                     NAME:PACKED ARRAY[1 .. 25] OF CHAR;
                     VALUE:REAL;
                     RATING:1 .. 100;
                     CASE SUPPLIER:AVAILABILITY OF
                     UK:
                     (ADDRESS:PACKED ARRAY[1 .. 50] OF CHAR;
                     PRICE:REAL;
                     DELIVERY:PACKED ARRAY[1 .. 10] OF CHAR);
                     NOTUK:
                     (COUNTRIES:PACKED ARRAY[1 .. 60] OF CHAR)
                     END;
```

We first define the enumerated type AVAILABILITY. In the record definition the three fixed fields are defined in the normal way. The statement CASE SUPPLIER:AVAILABILITY OF defines a special field SUPPLIER of type AVAILABILITY. This special field is known as the **tag field**. The variant part of the record is dependent on the value assigned to the tag field. Thus if the tag field SUPPLIER has the current value UK, then the 3 fields ADDRESS, PRICE, DELIVERY can be accessed; if SUPPLIER is set to NOTUK, then only the single field COUNTRIES can be accessed.

Example 7.3 This basic example illustrates the definition of the record COMPONENT consisting of 3 fields; the declaration of RESISTOR as an array of records of type COMPONENT; and the use of the WITH statement for the assignment and display of the first of the record array elements RESISTOR[1].

```
PROGRAM RECORD1;

TYPE COMPONENT = RECORD
                  NAME:ARRAY[1..20] OF CHAR;
                  VALUE:1..1000;
                  RATING:1..20
                END;

VAR RESISTOR : ARRAY[1..100] OF COMPONENT;

BEGIN
  WITH RESISTOR[1] DO
    BEGIN
      NAME:='WIRE-WOUND TYPE 10 W';
      VALUE:=470;
      RATING:=10
    END;
  WITH RESISTOR[1] DO
    BEGIN
      WRITELN(NAME,VALUE:8,RATING:8)
    END
END.
```

Example 7.4 This example illustrates the use of records in processing complex numbers. The record COMPLEXNO is defined to consist of 4 fields, REALPART, IMAGPART, MODULUS and ARGUMENT all of which are real. The program can be used to determine either

1 The polar form of a complex number when the real and imaginary parts are entered from the keyboard, or
2 The real and imaginary parts when the modulus and argument are entered.

```
PROGRAM COMPLEXNUMBERS;
{ To find x-y or polar form }
CONST    PI=3.14159265;

TYPE     COMPLEXNO = RECORD
                        REALPART:REAL;
                        IMAGPART:REAL;
                        MODULUS :REAL;
                        ARGUMENT:REAL
                     END;

VAR       Z                :COMPLEXNO;
          MODZ,ARGZ,X,Y    :REAL;
          N                :INTEGER;

PROCEDURE INPUTXY;
BEGIN
    WRITELN('ENTER REAL AND IMAG PARTS OF NUMBER');
    READ(Z.REALPART,Z.IMAGPART)
END;

PROCEDURE POLARFORM;
BEGIN
  WITH Z DO
   BEGIN
    MODZ:=SQRT(SQR(REALPART)+SQR(IMAGPART));
    IF (REALPART=0) AND (IMAGPART>0) THEN ARGZ:=90;
    IF (REALPART=0) AND (IMAGPART<0) THEN ARGZ:=-90;
    IF REALPART > 0 THEN
    ARGZ:=ARCTAN(IMAGPART/REALPART)*180/PI;
    IF REALPART < 0 THEN
    ARGZ:=180+ARCTAN(IMAGPART/REALPART)*180/PI
   END;
WRITELN('MODULUS OF COMPLEX NO.= ',MODZ:8:3);
WRITELN('ARGUMENT           = ',ARGZ:8:2)
END;

PROCEDURE INPUTRTHETA;
BEGIN
    WRITELN('ENTER MAGNITUDE AND ANGLE OF NUMBER');
    READ(Z.MODULUS,Z.ARGUMENT)
END;

PROCEDURE CARTESIANFORM;
BEGIN
  WITH Z DO
   BEGIN
    ARGUMENT:=ARGUMENT*PI/180;
    X:=MODULUS*COS(ARGUMENT);
    Y:=MODULUS*SIN(ARGUMENT)
   END;
WRITELN('REAL PART = ',X:8:4);
WRITELN('IMAG PART = ',Y:8:3)
END;

BEGIN  {MAIN PROGRAM}
WRITELN('1. TO FIND POLAR FORM OF COMLEX NUMBER');
WRITELN('GIVEN ITS REAL AND IMAG PARTS : ENTER 1');
WRITELN;
WRITELN('2. TO FIND POLAR FORM :          ENTER 2');
READ(N);
```

```
IF N=1 THEN    BEGIN
                  INPUTXY;
                  POLARFORM
               END;
IF N=2 THEN    BEGIN
                  INPUTRTHETA;
                  CARTESIANFORM
               END
END.
```

Example 7.5 The following program allows records of N students (e.g. their names, age, sex) to be entered in from the keyboard and stored in the array CLASS1. The maximum number is 50 in the program but this may obviously be changed by altering the dimension of CLASS1. Procedures ENTERRECORDS and READNAME accomplish the input tasks and can easily be modified to be able to accept the data if already on file.

The record data can be displayed by a call to the procedure DIS-PLAYRECORDS. The procedure SEARCH enables a search of the records to be made to find and display the record corresponding to a name entered in from the keyboard.

```
PROGRAM CLASSRECORDS;

TYPE   STRING = PACKED ARRAY[1..20] OF CHAR;
       STUDENT=RECORD
                  NAME:STRING;
                  AGE :16..70;
                  SEX :CHAR {* M or F *}
               END;
VAR    CLASS1: ARRAY[1..50] OF STUDENT;
       N:INTEGER;
       ANYNAME:STRING;

PROCEDURE READNAME(VAR SNAME:STRING);
VAR  I,LENG :INTEGER;
BEGIN
 I:=1;LENG:=0;
 READLN;
   WHILE NOT EOLN DO
     BEGIN
      READ(SNAME[I]);
      I:=I+1;LENG:=LENG+1
     END;
   IF LENG<20 THEN
     FOR I:=LENG+1 TO 20 DO
       SNAME[I]:=' ';
 READLN
END;
```

```
PROCEDURE SEARCH;
VAR K : INTEGER;
BEGIN
 READNAME(ANYNAME);
 K:=0;
   REPEAT
    K:=K+1
    UNTIL (CLASS1[K].NAME=ANYNAME) OR (K=N+1);
 IF K=N+1 THEN WRITE('NOT LISTED')
   ELSE
    WITH CLASS1[K] DO
     WRITELN(NAME,AGE:8,SEX:6)
END;

PROCEDURE ENTERRECORDS(N:INTEGER);
VAR K:INTEGER;
BEGIN
 FOR K:=1 TO N DO
  BEGIN
   WRITELN('ENTER NAME,AGE,SEX');
    WITH CLASS1[K] DO
     BEGIN
      READNAME(NAME);
      READLN(AGE);
      READLN(SEX)
     END
  END
END;

PROCEDURE DISPLAYRECORDS(N:INTEGER);
VAR  K:INTEGER;
BEGIN
 FOR K:=1 TO N DO
  BEGIN
   WITH CLASS1[K] DO
    WRITELN(NAME,AGE:8,SEX:6)
  END
END;

BEGIN  {*** MAIN PROGRAM ***}
WRITE('NUMBER OF STUDENTS = ');
READ(N);
ENTERRECORDS(N);
DISPLAYRECORDS(N);
WRITE('ENTER NAME FOR SEARCH: ');
READ(ANYNAME);
SEARCH
END.
```

7.4 Sets

PASCAL also provides us with the facility to operate with sets. In PASCAL, members of a set must be of the same ordinal type. All scalar types can be used except REAL.

Sets are defined as indicated in the following example:

```
TYPE    RESISTORS = (carbon,wire,metalfilm,metal-oxide);
        NODEPOSITIONS = 1 .. 20;
        SOURCEPOSITIONS = 'A'..'Z';
        RBOX = SET OF RESISTORS;
VAR     R1,R2:  RBOX;
        N1,N2,N3:   SET OF NODEPOSITIONS:
        G1,G2,G3:   SET OF SOURCEPOSITIONS;
```

In PASCAL, a set is built up by writing the data values for each member, separated by commas, and enclosed by rectangular brackets [ ]. Thus, for example, to assign values to a set variable we use statements of the form:

```
R1 := [carbon,wire,metal-oxide];
R2 := [wire,metalfilm,metal-oxide];
N1 := [1 .. 2];   N2 := [2 .. 3];
G1 := ['A'];   G2 := ['D'..'H'];
G3 := [ ] {*emptyset,containsnovalues*};
```

The following operations can be performed using the comparison operators which have the following meaning when used with sets:

Meaning	*PASCAL symbol*	*Example*
Set equality	=	R1 = R2
Set inequality	< >	R1 < > R2
Set "contains"	> =	R1 > = R2
Set "is contained in"	< =	R1 < = R2

and all yield Boolean results.

The union and intersection operators are represented in PASCAL using the symbols + and *, e.g. if

R1 := [carbon,wire,metalfilm]   R2 := [wire,metalfilm,metal-oxide]

then the **union** R1 + R2 yields the set containing all the members of both sets, i.e. [carbon, wire, metalfilm, metal-oxide].

The **intersection** R1 * R2 yields the set containing members that are common to both sets, i.e. [wire, metalfilm].

An additional operator **IN** is also available in PASCAL and is used to test for set membership or inclusion, e.g.

IF carbon IN R1 THEN WRITE('Carbon included');
IF LETTER IN['a','e','i','o','u']THEN WRITE('Vowel');

Note that the number of elements a set may contain is obviously limited and is typically 256. Note also that sets of INTEGERS defined *not* using the sub-range notation (e.g. 1..20) normally lead to problems, even with sophis-ticated compilers, since the cardinality of a set of integers exceeds 32 000 even for microcomputers. Thus, for example, assignments such as

N1:=[100,150,220]

cannot be used directly unless N1 has been previously declared as a set variable of a subrange type of, for example, N = 0..255.

Example 7.6 The following program illustrates an application of the IN operator to test for set membership and so find the preferred value of a resistor for values in the range 10 to 100 ohms. Note the values for ±20% tolerance resistors in this range are 10, 15, 22, 33, 47 68 and 100 Ω and such values are referred to as preferred values.

```
PROGRAM FINDPREFERREDVALUE;

VAR  R:INTEGER;

BEGIN
    WRITE('Enter value of resistor ');
    READ(R);
    IF R IN [10..12] THEN R:=10;
    IF R IN [13..18] THEN R:=15;
    IF R IN [19..26] THEN R:=22;
    IF R IN [27..39] THEN R:=33;
    IF R IN [40..57] THEN R:=47;
    IF R IN [58..84] THEN R:=68;
    IF R IN [85..100] THEN R:=100;
    WRITELN('PREFERRED VALUE =',R:6)
END.
```

Example 7.7 This simple program illustrates the definition of a set of a given enumerated type: the union (+) and intersection (*) operators used in assignment statements and subsequently the use of sets in the test condition in IF...THEN statements.

```
PROGRAM SET2;

TYPE  CAPACITORS=(PAPER,CERAMIC,MICA,TANTALUM);
      CBOX     = SET OF CAPACITORS;

VAR  C1,C2,C :CBOX ;

BEGIN
 C1:= [PAPER,CERAMIC];
 C2:= [PAPER,MICA,TANTALUM];
 C :=C1+C2 ;
 IF MICA IN C THEN WRITELN('MICA C FOUND');
 C:=C1*C2;
 IF C=[PAPER] THEN
   WRITELN('PAPER CAPACITORS IN BOTH BOXES')
END.
```

Exercises 7

7.1 Write a program that reads in names contained in an existing text file, sorts the names in alphabetic order, displays the sorted list, and then saves it on file.

7.2 Using the record data structure write procedures to a) add, b) subtract, c) multiply, d) divide two complex numbers.

7.3 Write procedures which will count the total number of characters and words in a text file.

7.4 A radar system has the following parameters:
wavelength $\lambda = 0\cdot03$ m; peak transmitter power $P_T = 80$ kW;
antenna gain $G = 500$; antenna noise temperature $T_S = 290$ K;
receiver noise temperature $T_R = 1200$ K; bandwidth $B = 5$ MHz;
signal-to-noise ratio for satisfactory reception $S_0/N_0 > 10$.

Express this information in the form of a record and use the record in a program designed to determine the maximum range of the radar for targets of various values of echoing area σ, given that the maximum range is

$$r_{max} = [\sigma G^2 \lambda^2 P_T/(64\pi^3 P_R)]^{0.25}$$

where

$$P_R = 1\cdot38 \times 10^{-23} \times B(T_S + T_R) \times S_0/N_0$$

ANSWERS TO EXERCISES

Exercises 2

2.1 The following are invalid identifiers:

1XRAY, X+Y, HI-TEMP, NO OF VALUES, 'ABC', +Z

Use the program of example 2.7 or the simple program given below:

```
PROGRAM AVERAGE;
VAR A,B,C,D,E,MEAN :REAL;
BEGIN
 WRITELN('Enter the 5 quantities separating each');
 WRITELN('each with a space');
 READ(A,B,C,D,E);
 MEAN:=(A+B+C+D+E)/5;
 WRITE('MEAN =',MEAN:10:1)
END.
```

2.4 A simple program to carry out the task is given below:

```
PROGRAM PARALLELR;
VAR R1,R2,G :REAL;
BEGIN
WRITE('ENTER RESISTOR VALUES,R1 AND R2: ');
READ(R1,R2);
G:=1/R1+1/R2;
WRITE('R= ',1/G:8:0)
END.
```

Exercises 3

3.1 Program to carry out calculations for *a*), *b*), *c*) and *d*):

```
PROGRAM CACULATIONS;
BEGIN
 WRITELN('ANS TO (a) :',SQRT(5.6*2.3));
 WRITELN('ANS TO (b) :',3.8*(5-0.65*2.1));
 WRITE('ANS TO (c) : ');
 WRITELN((5.6*2.3)/(10.2-9.4*0.7));
 WRITE('ANS TO(d): ');
 WRITELN(SQRT(69.8/SQR(12.67-2.98)))
END.
```

3.2 5; 3; 2

3.3 A typical program is

```
PROGRAM FEETTOMETRES;
VAR FEET,INCH :REAL;
BEGIN
 WRITE('ENTER FEET and INCHES');
 READ(FEET,INCH);
 WRITE(FEET:10:0,' FT',INCH:10:0,'IN =');
 WRITELN((12*FEET+INCH)*0.0254:10:4,' METRES')
END.
```

3.4 7·0; 14, −14; −15; 0·7853915; a; 63; C

3.5 Program to determine resistance:

```
PROGRAM RCALC;
VAR R1,R2,R3 :REAL;
BEGIN
 WRITE('ENTER R1,R2,R3 VALUES: ');
 READ(R1,R2,R3);
 WRITE('TOTAL R = ');
 WRITELN(R1+R2*R3/(R2+R3):10:2)
END.
```

3.6 Program to determine maximum and minimum resistance:

```
PROGRAM RCALC2;
VAR R1,R2,R3,R,T:REAL;
BEGIN
 WRITE('ENTER R1,R2,R3 VALUES ');
 READ(R1,R2,R3);
 WRITE('ENTER TOLERANCE %');
 READ(T);T:=T/100;
 R:=R1+R2*R3/(R2+R3);
 WRITELN('MIN VALUE OF R = ',R*(1-T):10:0);
 WRITELN('MAX VALUE OF R = ',R*(1+T):10:0)
END.
```

3.7 Program can take the typical form:

```
PROGRAM TRANSISTORBIAS;
CONST VBE=0.65;
VAR VCC,VCE,IC,HFE :REAL;
BEGIN
 WRITE('ENTER VALUE OF SUPPLY VOLTS: ');
 READ(VCC);
 WRITE('ENTER VALUE OF DC CURRENT GAIN: ');
 READ(HFE);
 WRITE('ENTER OPERATING POINT VALUES ');
 WRITE('VCE and IC ');
 READ(VCE,IC);
 WRITELN('RC = ',(VCC-VCE)/IC:8:0);
 WRITELN('RB = ',HFE*(VCC-VBE)/IC:8:0)
END.
```

3.8 A basic program for finding the transfer function:

```
PROGRAM TRANSFERFUNCTION;
CONST PI=3.14159;
VAR C,R,Z,F :REAL;
BEGIN
 WRITE('ENTER VALUES FOR C,R AND FREQUENCY');
 READ(C,R,F);
 Z:=SQRT(R*R+1/SQR(C*2*PI*F));
 WRITE('TRANSFER FUNCTION = ',R/Z:10:5)
END.
```

3.9 There are a number of ways of doing this. The program below uses the FOR loop (see Chapter 4).

```
PROGRAM RESPONSE1;
CONST PI=3.14159;
VAR  N:INTEGER;
     C,R,F,Z:REAL;
BEGIN
 WRITE('ENTER VALUES FOR C AND R');
 READ(C,R);
 F:=1;
 WRITELN('FREQUENCY':10,'TRANSFER FUNC':25);
 FOR N:=1 TO 6 DO
  BEGIN
   F:=F*10;
   Z:=SQRT(R*R+1/SQR(C*2*PI*F));
   WRITELN(F:10:0,R/Z:20:5)
  END
END.
```

3.10 Suitable program for determining transmission loss:

```
PROGRAM LINKLOSS;
CONST PI=3.14159;C=3.0E08;
VAR F,GT,GR,PT,R,PR :REAL;
BEGIN
 WRITELN('FREQUENCY = ');READ(F);
 WRITE('GAINS OF TRANSM AND REC ANTENNAS');
 READ(GT,GR);
 WRITE('TRANSMITTER POWER = ');READ(PT);
 WRITE('RANGE = ');READ(R);
 PR:=SQR(C/F)*GT*GR*PT/SQR(4*PI*R);
 WRITELN('RECEIVED POWER = ',PR,' WATTS');
 WRITE('TRANSMISSION LOSS = ');
 WRITE(4.343*LN(PT/PR):8:0,' dB')
END.
```

Note: Log to base 10 is not a standard PASCAL function; use

$$\log_{10} = 4 \cdot 343 \ln.$$

Exercises 4

4.1 *a*) A suitable program:

```
PROGRAM GREATER;
VAR A,B :REAL;
BEGIN
 WRITE('ENTER A and B: ');
 READ(A,B);
 IF A>B THEN
  WRITELN(A:10:2,' IS GREATER')
 ELSE
  WRITELN(B:10:2,' IS GREATER')
END.
```

b) To find maximum and minimum values:

```
PROGRAM MAXMIN;
VAR N :INTEGER;
    X,NMAX,NMIN :REAL;
BEGIN
 NMAX:=-10E10;NMIN:=10E10;
 FOR N:=1 TO 20 DO
  BEGIN
   WRITE('ENTER VALUE: ');
   READ(X);
   IF X>=NMAX THEN NMAX:=X;
   IF X<=NMIN THEN NMIN:=X
  END;
WRITELN('MAX = ',NMAX:10:2);
WRITELN('MIN = ',NMIN:10:2)
END.
```

4.2 There are many different approaches; here is one solution:

```
PROGRAM  MONITOR;
VAR V1,V2,V3 :REAL;
    V1COND,V2COND,V3COND,TOTALCOND :BOOLEAN;
BEGIN
 TOTALCOND:=TRUE;
 WHILE TOTALCOND=TRUE DO
 BEGIN
  WRITE('ENTER V1,V2,V3: ');
  READ(V1,V2,V3);
  V1COND:=(V1>=0) AND (V1<=1);
  V2COND:=(V2>=10) AND (V2<=12);
  V3COND:=(V3>=100) AND (V3<=120);
  TOTALCOND:=V1COND AND V2COND AND V3COND
 END;
IF V1COND=FALSE THEN WRITELN('V1= ',V1:6:1);
IF V2COND=FALSE THEN WRITELN('V2= ',V2:6:1);
IF V3COND=FALSE THEN WRITELN('V3= ',V3:6:1)
END.
```

4.3 *a*) A program to clear the screen:

```
PROGRAM CLEARSCREEN;
VAR N:INTEGER;
BEGIN
FOR N:=1 TO 25 DO
  WRITELN
END.
```

b) A program to produce a delay:

```
PROGRAM DELAY;
VAR N:INTEGER;
BEGIN
 WRITELN('VALUE OF N DETERMINES THE DELAY');
 WRITELN('DELAY BEGINS');
  FOR N:=1 TO 10000 DO
   {* NOTHING *};
 WRITELN('DELAY ENDS')
END.
```

c) A program to produce the table of values:

```
PROGRAM TABLEDISPLAY;
VAR N:INTEGER;
    X:REAL;
BEGIN
 FOR N:=0 TO 20 DO
 BEGIN
  X:=0.2*N;
  WRITELN(X:5:2,EXP(-X):10:6,(1-EXP(-X)):10:6)
 END
END.
```

4.4 The core of this program is typically:

```
PROGRAM THERMOSIMULATOR;
VAR TEMP:REAL;
BEGIN
WRITE('ENTER TEMP: ');
READ(TEMP);
IF (TEMP<=15) THEN
 WRITELN('SWITCH ON HEATER+BOOST');
IF (TEMP>15) AND (TEMP<22) THEN
 WRITELN('HEATER ON ONLY');
IF (TEMP>22) AND (TEMP<26) THEN
 WRITELN('SWITCH OFF HEATER');
IF TEMP>=26 THEN
 WRITELN('SWITCH ON FAN')
END.
```

4.5 Program to find mean and rms for a number of discrete values:

```
PROGRAM MEANANDRMS;
VAR X,SUM,SUMSQ   :REAL;
    N:INTEGER;
BEGIN
 SUM:=0;SUMSQ:=0;N:=0;
  REPEAT
   WRITE('ENTER X: ');
   READ(X);
   N:=N+1;
   SUM:=SUM+X;
   SUMSQ:=SUMSQ+X*X;
  UNTIL X=0;
WRITELN('MEAN = ',SUM/(N-1));
WRITELN('RMS  = ',SQRT(SUMSQ/(N-1)))
END.
```

4.6 *a*) This is an approximate solution where $X = \sin(\pi N/20)$ is essentially divided into 21 discrete values between 0 and π. These values are summed and divided by 41 to account for the fact that the half sinewave is zero between 0 and 2π.

```
PROGRAM SINEAVERAGE;
CONST PI=3.14159;
VAR X,SUM,SUMSQ:REAL;
    N: INTEGER;
BEGIN
  SUM:=0;SUMSQ:=0;N:=0;
   REPEAT
    X:=SIN(PI*N/20);
    SUM:=SUM + X;
    SUMSQ:=SUMSQ+X*X;
    N:=N+1
   UNTIL N=20;
WRITELN('MEAN = ',SUM/41:6:2);
WRITELN('RMS  = ',SQRT(SUMSQ/41):6:2);
WRITELN('        APPROXIMATELY')
END.
```

   b) Use approach similar to a).

4.7 Program to determine normalized frequency response of a tuned circuit:

```
PROGRAM TUNEDCICUITRESPONSE;
CONST STEP=0.02;
VAR V,D,Q  :REAL;
    N :INTEGER;
BEGIN
 WRITE('INPUT Q-VALUE: ');
 READ(Q);
 D:=-0.2;
 FOR N:=1 TO 20 DO
  BEGIN
   V:=1/SQRT(1+SQR(2*D*Q));
   WRITELN(D:6:2,V:10:5);
   D:=-0.2+STEP*N
  END
END.
```

Exercises 5

5.2 A suitable program containing the two functions:
```
PROGRAM  RCALC;
VAR R1,R2,R3,RS,RP :REAL;

FUNCTION RSERIES(RA,RB,RC:REAL):REAL;
BEGIN
 RSERIES:=RA+RB+RC
END;

FUNCTION RPARAL(RA,RB,RC:REAL):REAL;
BEGIN
 RPARAL:=1/(1/RA+1/RB+1/RC)
END;

BEGIN {* MAIN PROGRAM *}
WRITE('ENTER VALUES OF 3 PARAL. ELEMENTS: ');
READ(R1,R2,R3);
RP:=RPARAL(R1,R2,R3);
WRITE('ENTER VALUES OF 2 SERIES ELEMENTS: ');
READ(R1,R2);
RS:=RSERIES(R1,R2,RP);
WRITE(' TOTAL R = ');
WRITELN(RS:8:1,' OHMS')
END.
```

5.3 Procedure QUADROOTS takes into account both real and complex roots:

```
PROGRAM QUADSOLUTION;
VAR A,B,C,X1,X2 :REAL;

PROCEDURE QUADROOTS(A,B,C:REAL);
VAR D:REAL;
BEGIN
 D:=B*B-4*A*C;
 IF D>=0 THEN
  BEGIN
   X1:=(-B+SQRT(D))/(2*A);
   X2:=(-B-SQRT(D))/(2*A);;
   WRITELN('ROOTS ARE: ');
   WRITELN(X1:8:3,X2:8:3)
  END
 ELSE
  BEGIN
   WRITELN('ROOTS ARE COMPLEX:');
   WRITELN(-B/(2*A):8:3,' +j ',SQRT(-D)/(2*A):8:3);
   WRITELN(-B/(2*A):8:3,' -j ',SQRT(-D)/(2*A):8:3)
  END
END;

BEGIN {* MAIN PROGRAM *}
 WRITE('ENTER a,b,c COEFFICIENTS OF EQUATION: ');
 READ(A,B,C);
 QUADROOTS(A,B,C)
END.
```

5.4 The four functions are given in the following program.
Note: tan (x) is not a standard function in many PASCAL implementations.

```
PROGRAM DIFFERENTFUNCTIONS;
VAR X,Y,Z :REAL;

FUNCTION AVERAGE(A,B,C:REAL):REAL;
BEGIN
AVERAGE:=(A+B+C)/3
END;

FUNCTION MAX(A,B,C:REAL):REAL;
VAR M:REAL;
BEGIN
 IF A>=B THEN M:=A
 ELSE M:=B;
 IF C>=M THEN M:=C;
 MAX:=M
END;

FUNCTION MIN(A,B,C:REAL):REAL;
VAR M:REAL;
BEGIN
 IF A<=B THEN M:=A
 ELSE M:=B;
 IF C<=M THEN M:=C;
 MIN:=M
END;
```

```
FUNCTION TAN(A:REAL):REAL;
BEGIN
TAN:=SIN(A)/COS(A)
END;

BEGIN {* MAIN PROGRAM *}
WRITE('ENTER 3 NOS: ');
READ(X,Y,Z);
WRITELN('MAX = ',MAX(X,Y,Z):8:3);
WRITELN('MIN = ',MIN(X,Y,Z):8:3);
WRITELN('AVERAGE = ',AVERAGE:8:3);
WRITELN('TAN(A RADS) =',TAN(X):8:3)
END.
```

5.5 *a*) Program for finding series impedance of $Z_1 = R_1 + jX_1$ and $Z_2 = R_2 + jX_2$:

```
PROGRAM SERIESIMPEDANCE;
VAR R1,R2,X1,X2,RT,XT:REAL;
PROCEDURE ZS(R1,X1,R2,X2:REAL;VAR RS,XS:REAL);
BEGIN
RS:=R1+R2;
XS:=X1+X2
END;

BEGIN {* MAIN PROGRAM *}
WRITE('ENTER RESISTIVE COMP. OF Z1: ');
READ(R1);
WRITE('ENTER REACTIVE COMP. OF Z1: ');
READ(X1);
WRITE('ENTER R AND W COMPS. OF Z2: ');
READ(R2,X2);
ZS(R1,X1,R2,X2,RT,XT);
WRITE('TOTAL Z = ');
WRITELN(RT:8:3,' +j ',XT:8:3);
END.
```

b) Program for finding impedance of two impedances in parallel:

```
PROGRAM PARALLELIMPEDANCE;
VAR R1,X1,R2,X2,RT,XT :REAL;

  PROCEDURE ADMITTANCE(R,X:REAL;VAR G,B:REAL);
   BEGIN
    G:=R/(R*R+X*X);
    B:=-X/(R*R+X*X)
   END;

PROCEDURE ZP(R1,X1,R2,X2:REAL;VAR RT,XT:REAL);
VAR G1,B1,G2,B2,GT,BT:REAL;
 BEGIN
  ADMITTANCE(R1,X1,G1,B1);
  ADMITTANCE(R2,X2,G2,B2);
  GT:=G1+G2;
  BT:=B1+B2;
  ADMITTANCE(GT,BT,RT,XT)
END;

BEGIN {* MAIN PROGRAM *}
WRITE('ENTER R AND X COMPS. OF Z1: ');
READ(R1,X1);
WRITE('ENTER R AND X COMPS. OF Z2: ');
READ(R2,X2);
ZP(R1,X1,R2,X2,RT,XT);
WRITE('TOTAL Z = ');
WRITELN(RT:8:4,' +j ',XT:8:4)
END.
```

5.6 Program to determine frequency response of filter section:

```
PROGRAM FRESPONSE;
VAR R,L,C,F:REAL;
    N:INTEGER;
FUNCTION VO(R,L,C,F:REAL):REAL;
VAR W:REAL;
BEGIN
W:=2*3.14159*F;
VO:=R/SQRT(SQR(R*(1-W*W*L*C))+SQR(L*W*(2-W*W*L*C)))
END;

BEGIN
WRITE('ENTER VALUES FOR R,L,C');
READ(R,L,C);
WRITELN('F':8,'VO':9);
FOR N:=1 TO 20 DO
  BEGIN
  F:=1000*N;
  WRITELN(F:8:0,VO(R,L,C,F):12:4)
  END
END.
```

5.8 Program containing procedure SIMPSONRULE to effect numerical integration:

```
PROGRAM INTEGRATION1;
VAR LOWLIMIT,UPLIMIT,INTEGRAL:REAL;
    NSTEP  :INTEGER;

PROCEDURE SIMPSONRULE(A,B:REAL;NOSTEPS:INTEGER;VAR AREA:REAL);
VAR H,SUMOFY:REAL;
    N:INTEGER;
FUNCTION Y(X:REAL):REAL;
BEGIN
  Y:=2*X*X*X-21*X*X+60*X
END;
  BEGIN
  H:=(B-A)/NOSTEPS;
  SUMOFY:=Y(A)+Y(B);
    FOR N:=1 TO NOSTEPS-1 DO
      BEGIN
        IF ODD(N) THEN
          SUMOFY:=SUMOFY+4*Y(A+N*H)
        ELSE
          SUMOFY:=SUMOFY+2*Y(A+N*H)
      END;
AREA:=SUMOFY*H/3
END;

BEGIN {* MAIN PROGRAM *}
WRITE('ENTER LOWER AND UPPER LIMITS: ');
READ(LOWLIMIT,UPLIMIT);
WRITE('ENTER NO. OF STEPS ');
READ(NSTEP);
SIMPSONRULE(LOWLIMIT,UPLIMIT,NSTEP,INTEGRAL);
WRITELN('AREA = ',INTEGRAL)
END.
```

5.9 and **5.10** Functions for calculation of Z_0, α and β are included in the program, which can also be utilized for **5.10**.

```
PROGRAM TRANSMISSIONLINE1;
CONST PI=3.14159;
      RT=1.414;{SQ.ROOT OF 2}
VAR R,L,C,G,F :REAL;

FUNCTION ZO(R,L,C,G,F:REAL):REAL;
VAR W,ZS,YS :REAL;
BEGIN
 W:=2*PI*F;
 ZS:=R*R+W*W*L*L;  YS:=G*G+W*W*C*C;
 ZO:=SQRT(SQRT(ZS/YS))
END;

FUNCTION ALPHA(R,L,C,G,F:REAL):REAL;
VAR W,ZS,YS,ZY,K :REAL;
BEGIN
 W:=2*PI*F;
 ZS:=R*R+W*W*L*L;YS:=G*G+W*W*C*C;
 ZY:=SQRT(ZS*YS);K:=R*G-W*W*L*C;
 ALPHA:=SQRT(K+ZY)/RT
END;

FUNCTION BETA(R,L,C,G,F:REAL):REAL;
VAR W,ZS,YS,ZY,K:REAL;
BEGIN
W:=2*PI*F;
ZS:=R*R+W*W*L*L;YS:=G*G+W*W*C*C;
ZY:=SQRT(ZS*YS);K:=R*G-W*W*L*C;
BETA:=SQRT(-K+ZY)/RT
END;
 BEGIN {* MMAIN PROGRAM *}
 WRITE('ENTER LINE CONSTANTS: R,L,C,G: ');
 READ(R,L,C,G);
 WRITE('ENTER FREQUENCY: ');
 READ(F);
 WRITELN('AT F = ':15,F:9:0);
 WRITELN('ZO =    ':15,ZO(R,L,C,G,F):9:2);
 WRITELN('ATTENUATION = ':15,ALPHA(R,L,C,G,F):9:5);
 WRITELN('PHASE-CONSTANT = ':20,BETA(R,L,C,G,F):9:5)
END.
```

Exercises 6

6.1 A suitable program:

```
PROGRAM ENTERANDDISPLAY;
VAR NUMBER : ARRAY[1..20] OF INTEGER;
    COST   : ARRAY[1..20] OF REAL;

PROCEDURE ENTERDATA;
VAR J:INTEGER;
BEGIN
 FOR J:=1 TO 20 DO
 BEGIN
  WRITE('ENTER NO. AND UNIT-COST: ');
  READ(NUMBER[J],COST[J])
 END
END;

PROCEDURE DISPLAYTABLE;
VAR J:INTEGER;
BEGIN
 WRITELN('NUMBER':8,'UNIT-COST':10,'SUB-TOTAL':12);
  FOR J:=1 TO 20 DO
    WRITELN(NUMBER[J]:8,COST[J]:10:2,NUMBER[J]*COST[J]:12:2)
END;


BEGIN {* MAIN PROGRAM *}
 ENTERDATA;
 DISPLAYTABLE
END.
```

6.2 Program which reads 32 rows × 4 columns of data, calculate and displays averages:

```
PROGRAM ARRAYAVERAGE;
VAR ITEM :ARRAY[1..32,1..4] OF REAL;
    ROWAV:ARRAY[1..32] OF REAL;
    COLAV:ARRAY[1..4] OF REAL;
    ROW,COL :INTEGER;

PROCEDURE ENTERIN;
 BEGIN
  FOR ROW:=1 TO 32 DO
   FOR COL:= 1 TO 4 DO
    BEGIN
     WRITE('ENTER ITEM VALUE: ');
     WRITE(ROW,COL);
     READ(ITEM[ROW,COL])
    END
 END;

PROCEDURE ROWAVERAGE;
VAR SUM :REAL;
BEGIN
  FOR ROW:=1 TO 32 DO
   BEGIN
    SUM:=0;
    FOR COL:=1 TO 4 DO
     SUM:=SUM+ITEM[ROW,COL];
    ROWAV[ROW]:=SUM/4
   END
END;

PROCEDURE COLAVERAGE;
VAR SUM :REAL;
BEGIN
 FOR COL:=1 TO 4 DO
  BEGIN
   SUM:=0;
   FOR ROW:=1 TO 32 DO
    SUM:=SUM+ITEM[ROW,COL];
   COLAV[COL]:=SUM/32
  END
END;

PROCEDURE DISPLAY;
BEGIN
 FOR ROW:=1 TO 32 DO
 BEGIN
  WRITE(ROW:3);
   FOR COL:=1 TO 4 DO
   WRITELN( ITEM[ROW,COL]:5:1,ROWAV[ROW]:5:1);
 END;
WRITELN;
FOR COL:=1 TO 4 DO
WRITE('    ',COLAV[COL]:5:1)
END;

BEGIN {* MAIN PROGRAM *}
 ENTERIN;
 ROWAVERAGE;
 COLAVERAGE;
 DISPLAY
END.
```

6.3 Construct individual procedures for AVERAGE, RMS, MAXMIN and SORT, all of which are given in some form in the text. Construct also a procedure for entering data and counting number of terms, the data being stored in an array. Use the CASE statement in the main program to select your procedure

6.4 Use approach directly similar to Example 6.6.

6.5 Amongst many different ways of accomplishing the task, one simple solution for *a*) and *b*):

```
PROGRAM FILETASK1;

VAR    DATAFILE:FILE OF REAL;
       N         : INTEGER;

PROCEDURE ENTERANDCREATEFILE;
VAR    X : REAL;
BEGIN
  REWRITE(DATAFILE,'B:DATAFILE.001');
    FOR N:=1 TO 24 DO
      BEGIN
        WRITE('ENTER VALUE: ');
        READ(X);
        WRITE(DATAFILE,X)
      END
END;

PROCEDURE DISPLAYETC;
VAR    SUM,MAX,MIN,X:REAL;
BEGIN
  RESET(DATAFILE,'B:DATAFILE.001');
  MAX:=-10E10;MIN:=10E10;SUM:=0;
    FOR N:=1 TO 24 DO
      BEGIN
        READ(DATAFILE,X);
        SUM:=SUM + X;WRITELN(X);
        IF X>MAX THEN MAX:=X;
        IF X<MIN THEN MIN:=X
      END;
  WRITELN('AVERAGE =',SUM/N:10:3);
  WRITELN('MAXIMUM =',MAX:10:2);
  WRITELN('MINIMUM =',MIN:10:2)
END;

BEGIN    {*** MMAIN PROGRAM ***}
  ENTERANDCREATE;
  DISPLAYETC
END.
```

6.6 Suppose the external file containing the integer data is located on disc in disc drive B and has the filename NUMBERS1.INT. The following program reads in the data and sorts it into 7 bands.

```
PROGRAM FILETASK2;

VAR  NUMBERS1  :FILE OF INTEGER;
     C  :ARRAY[1..7] OF INTEGER;

PROCEDURE INITIALIZECOUNT;
VAR  M  :INTEGER;
BEGIN
  FOR M = 1 TO 7 DO
    C[M] := 0
END;

PROCEDURE READANDSORT;
VAR N,T  :INTEGER;
BEGIN
  RESET(NUMBER1,'B:NUMBERS1.INT');
  WHILE NOT EOF(NUMBERS1) DO
    BEGIN
     READ(NUMBERS1,N);
     IF N<=20 THEN C[1]:=C[1]+1;
     IF (N>20) AND (N<=30) THEN C[2]:=C[2]+1;
     IF (N>30) AND (N<=40) THEN C[3]:=C[3]+1;
     IF (N>40) AND (N<=50) THEN C[4]:=C[4]+1;
     IF (N>50) AND (N<=60) THEN C[5]:=C[5]+1;
     IF (N>60) AND (N<=70) THEN C[6]:=C[6]+1
    END;
  FOR T := 1 TO 7 DO
    WRITELN('COUNT RANGE',T:3,' = ',C[T]:8)
END;

BEGIN   {*** MAIN PROGRAM ***}
  INITIALIZECOUNT;
  READANDSORT
END.
```

8.6 Suppose the external file containing the telephone data is located on disc in disc drive 3 and has the filename 'PHONRES.DAT', the following program reads in the data and sorts it into T names.

```
PROGRAM TELETON

VAR NUMERAL : ARRAY OF INTEGER;
    D : GENVALL, VLIST : INTEGER;

PROCEDURE INITIALTECHLITES;
    VAR N : INTEGER;
    BEGIN
      FOR INDEX TO TEST;
    END;

PROCEDURE READTOFFILE;
    VAR N : INTEGER;
    BEGIN
      RESET (ZIP, VLIST);
      READ (NUMBER, DIMEBRGELMST);
      VAR K, RT, RGH NUMBERS, US;
    BEGIN
      BEGIN OF READ
      K := TO THEN OR THELLIT LIST;
      IF NUMBER = TO (VEST) THEN USE(RT) VLIST
      IF GOTO VALUALL +10 THEN CIT;DLOOP
      READ N, RD, LAND DREADS THEN READ(X)U=
      IF GOTO,TH NH=RD=TR;I DR(X)=TR;I +
      WHILE(NUT AND BTR = R THEN TRUE;
    END;

    WRITELN LISTED
    WRITELN (RESPONSE,RGOE,TEST)=.01LIST
    END;

BEGIN (MAIN PROGRAM)
    INITIALTECHLITES;
    READTOFFILE;
    END.
```

Index